Gustave Guttenberg

Botanical Guide through the Phipps Conservatories

In Pittsburg and Allegheny

Gustave Guttenberg

Botanical Guide through the Phipps Conservatories
In Pittsburg and Allegheny

ISBN/EAN: 9783337212100

Printed in Europe, USA, Canada, Australia, Japan

Cover: Foto ©berggeist007 / pixelio.de

More available books at **www.hansebooks.com**

BOTANICAL GUIDE

THROUGH THE

PHIPPS CONSERVATORIES

. . IN . .

PITTSBURG AND ALLEGHENY.

PREPARED BY

GUSTAVE GUTTENBERG,

TEACHER OF BIOLOGY, CENTRAL HIGH SCHOOL,

PITTSBURG, PA.

CONTENTS.

   I. BUILDINGS AND MANAGEMENT.
  II. THE PALM HOUSES.
 III. FOLIAGE PLANTS.
  IV. ORCHIDS.
   V. THE FLOWERS.
  VI. CLIMBING PLANTS.
 VII. AQUATIC PLANTS.
VIII. CACTI.
  IX. CURIOS.
   X. FERNS.
  XI. CLOSING REMARKS.
         The Guide.
         The Florists.
         The Parks.
         Proposed Extension of Schenley Park Conservatory.
         Botanical Gardens.
 XII. ALPHABETICAL INDEX.

A BEAUTIFUL MEMORIAL WINDOW EXECUTED BY

PETGEN BROS. ART GLASS WORKS,

LEADED
STAINED AND MOSAIC GLASS OF EVERY
DESCRIPTION,
FOR CHURCHES AND DWELLINGS,

OFFICE AND WORKS,
HAMILTON AND FIFTH AVES. E. E.
TELEPHONE EAST END, 326,
PITTSBURGH, PA.

PREFACE.

SINCE the Conservatory in Allegheny Park was inaugurated, seven years ago, it has been visited by thousands every week ; and since the SCHENLEY PARK CONSERVATORY has been thrown open to the public, it is visited by tens of thousands. Indeed the generous and public spirited citizen, MR. HENRY PHIPPS, JR., to whom the two cities owe these institutions, may justly feel gratified to see his gifts to the people so immediately and so universally appreciated. The fact that the two floral palaces have become so great favorites of the people, shows that the love of the beautiful is not wanting among them ; and this love is further fostered and developed by the floral feasts prepared with rare taste and skill by those who have charge of the institutions.

The author of this little book has been a frequent and delighted visitor to the Conservatories, which offer such excellent chance to study the children of nature from the most favored climes of our earth. Many a time when he was taking notes about some rare or interesting specimen, he soon found himself surrounded by a group of visitors, from whose questions, guesses and suggestions, he noticed the eager desire to know something more than the name of the admired plants.

This fact and the wish to avail himself of the Conservatories, as a help in teaching his classes in Botany, induced the author to prepare this guide. He is under great obligations to the Superindents of the Conservatories and their assistants ; he had free access to the valuable library of Mr. Hamilton, as well as to the books of reference in Mr. Bennett's office. The assistants helped him in getting full lists of the plants and in making arrangements for photographing groups of plants and single specimens.

The books principally consulted for information were "Nicholson's Dictionary," "The Treasury of Botany," "Johnson's Gardener's Dictionary," and "Orchid Growers' Manual"

As a first attempt in this direction, this book may have its shortcomings; the author will be glad to receive suggestions as to how it can be improved. Meanwhile this guide is sent out with the author's wish and hope that it may find favor with the public; that it may increase the interest in the plants cultivated in the Conservatories, and in flower culture in general; that it may add to the delight in beauty, the pleasures which closer acquaintance gives.

GUSTAVE GUTTENBERG.

PITTSBURG, PA., May, 1894.

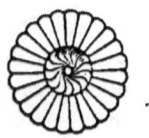

THE PHIPPS CONSERVATORIES.

BUILDINGS . .
AND MANAGEMENT.

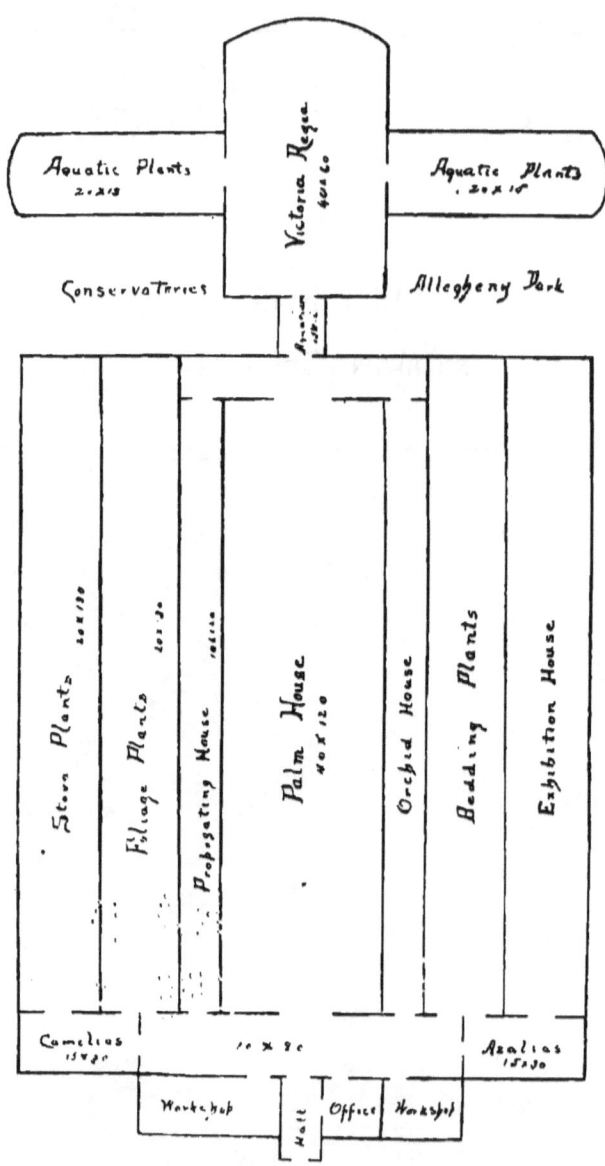

Plan of Allegheny Conservatory.

ALLEGHENY CONSERVATORY.

IN 1886, MR. HENRY PHIPPS, JR., of Allegheny, Pa., proposed to build, at his expense, a Conservatory in the Allegheny Parks, and to present it as a gift to the city, (if the latter would provide for its management, and under the condition that it should be open to the public on every day during the week, including Sundays.) This condition caused, at first, some opposition, but it was overcome, fortunately, for experience has shown since, that the closing on Sundays would have shut out the very people for whose benefit the gift was principally intended—the working-men and their families, who now are crowding the Conservatory every Sabbath-day afternoons.

A Committee, consisting of Mr. Wm. Hamilton, John Walker and O. P. Scaife, were appointed by Mr. Phipps to act as trustees. The foundations of the structure were laid in 1886, and the main buildings were finished in 1887; two years later the aquatic annex was added. The cost of the main buildings were $35,000, and of the aquatic annex $15,000.

As soon as finished the Conservatory was stocked with many fine and rare plants, a number of citizens of Allegheny having subscribed to a fund for this purpose. The new institution thus handsomely started, was placed in charge of Mr. William Hamilton, Superintendent of the Allegheny Parks.

The buildings consist of a Palm House 40 x 120 feet; parallel with this is the Orchid House, which contains a large and fine collection of Orchids, also Ferns and Pitcher Plants. Next to the Orchid house is a house for the care of the bedding plants, which are set out in Summer to adorn the Park. The last to the right is the Exhibition house, where Mr. Hamilton arranges his fine displays of Chinese Primroses, Cinerarias, Calceolari'as, Chrysanthemums, etc. To enter this house, the visitor has to pass the "Azalia corner," where for years a profusion of Azalias have greeted the visitor during part of April and May. To the left of the Palm House is the propagating house, where thousands of plants, intended for bedding or for exhibits are reared from the seed or slips and cuttings. To the left of this are two more houses filled with Cacti, foliage plants and other tropical or so-called stove plants, such as Begonias, Dracænas, Crotons, Euphorbias, and many others. The way to the last of these houses leads through an ante-room, where a group of Camelias is generally found, together with some Laurel trees, Agaves, Strelitzias and other plants. From the rear of the Palm house one enters into the Aquarium, where curious, interesting fish can be seen when the water happens to be clear. Next we pass into the large Victoria Regia house, from which extend two side wings devoted to the rearing of rare aquatic plants.

The dimensions of all of these houses can be found in the adjoining plan.

As mentioned above, the Conservatory is under the management of *Mr. William Hamilton;* his assistants are *Mr. John Herron*, foreman; *Mr. James C. Hamilton* has charge of the Orchids; *Mr. John Small* of the "Stove" plants, and *Mr. William Gibbs* of the Temperate houses. They all attend to their special duties with that love and devotion which characterizes the true gardener, who has made the care of plants his life calling.

SUPERINTENDENT HAMILTON AND HIS ASSISTANTS.

Do you Wish to Buy a Piano or Organ?

Or anything in the Musical line? We have beautiful Parlor Organs in Oak or Walnut,

At $38, $45, $55, $75, &c.

All fully warranted, for house, church or chapel use, and Upright Pianos of well known makes,

At $200, $250, $275, $300, &c.

We also have genuine Pipe Organs erected on salesroom floor, where you can see and hear them. In no other place in the city will you find such a selection of

Pianos, Organs and Musical Instruments.

Easy terms of payment will be made. It will pay you for your house, your churches and your schools to come in and see us or write to us for prices and terms.

~ S. HAMILTON, ~

91 & 93 Fifth Ave.   PITTSBURG, PA.

YOU SHOULD Have a Microscope when you examine Flowers, Plants, etc., and see some of their hidden beauty. We have them from 25 cents and upwards.

WM. E. STIEREN, OPTICIAN,

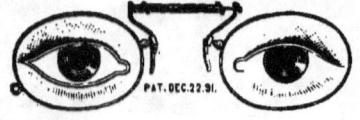

544 Smithfield St.   :   PITTSBURG, PA.

ESTABLISHED 1872.

J. L. McSHANE & CO.

SANITARY PLUMBERS AND GAS FITTERS,

19 SEVENTH AVENUE,

Pumps of all kinds. Gas and Electric Fixtures of the Latest Designs. Telephone 18.   PITTSBURG, PA.

FINE PLANTS.   CHOICE FLOWERS.

Randolph & McClements,
Florists,

Cor. South Highland Ave. and Baum St.

East End,

TELEPHONE E. E. 25.   Pittsburgh, Pa.

Areca and Fan Palms.

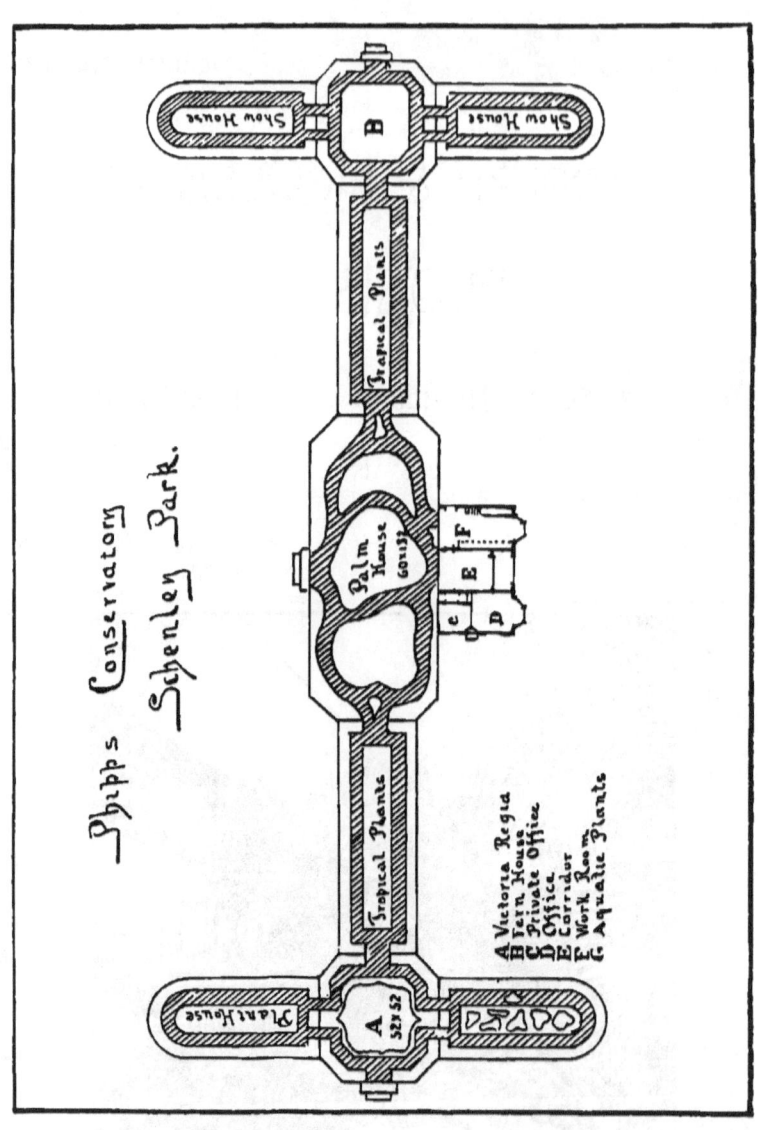

Schenley Park Conservatory.

IT was probably the great success which the establishment of the Conservatory in Allegheny had proved to be, and the popularity it enjoyed from the beginning, that prompted Mr. Phipps to make even a more generous offer to the city of Pittsburg. He offered to donate $100,000 towards the erection of a Conservatory in Schenley Park, and placed the matter into the hands of Mr. E. M. Bigelow, Chief of Department of Public Works ; Mr. John Walker and Mr. O. P. Scaife.

The offer was accepted by the city, and the people of Pittsburg enjoy to-day the possession of the largest and handsomest public Conservatory in this country, if not in the world.

The plan of the building as it stands to-day can be seen in the adjoining illustration. The total frontage is 454 feet. The centre is occupied by a Palm House, 60 feet wide and 132 feet long, with an elegant domed roof 60 feet high ; to the right and left of this house are wings 30 feet wide by 104 feet long ; these are intended for so-called " stove " plants; these in horticultural language, are plants which need a rather high temperature for their successful raising ; this temperature was in former times obtained by having a stove in the greenhouse. At present the necessary warmth is obtained much more conveniently and effectually by a steam heating apparatus. The unpoetic name of " stove house " or " stove plants " might therefore be now discarded and the attempt is made in this book to introduce instead, the name of " Tropical plants."

At present the right wing contains fine collections of foliage plants, Begonias, Orchids and other flowering plants and Ferns. A part of a section in both wings is devoted to propagating purposes, and to see the tiny Gloxinias, Cyclamens, etc., start from the seed, and the Ferns from the " spawn," Begonias, Cacti, growing from a leaf placed upon the sand, the Coleus, Fuchsias, Mesembryanthemums and others, raised from slips, is not less interesting to many visitors than to see the plants in the glory of their bloom.

The left wing contains now a collection of Cactus plants and a great mass of bedding plants tastefully interrupted by Passifloras, Aristolochias and other climbers or shrubby plants. Later this wing will be devoted to the culture of Orchids and other tropical plants.

Each of these wings leads into an octagonal domed-roofed building, 52 feet in diameter. The one adjoining the right or north-western wing contains the famous Tree Ferns, Staghorn Ferns, Birds-nest and other Ferns and some Araucarias, the other, adjoining the left or south-eastern wing, contains the tank for the monstrous water plant Victoria Regia. From each end of the octagons again run two side-wings at right angles to the main wings, each of them is 84 feet long and 32 feet wide. The wings right and left from the Fern house are devoted to the floral exhibits which prove such an attractive part of the Conservatory. The wing to the left of the Victoria house is used at present as a nursery for the plants to be exhibited when their time for blooming arrives; the other is arranged with cemented basins and metal tanks for the rearing of aquatic plants.

In the front of the Palm house is a fire-proof building 32 feet long and 53 feet wide, divided into two parts, by the corridor. At the left of the entrance are the offices of the Superintendent and at the right the working rooms of the gardeners. The corridor is guarded by the huge and handsome dogs of so-called green marble, really Serpentine; they are the work of an Italian sculptor, and have been presented by Mr. Phipps, who also donated the "Kneeling Venus," of white marble, who, in her verdant frame, is so greatly admired by the visitors.

The Conservatory buildings cover an area of about 34,000 feet of ground. The glass surface is over 60,000 square feet. It is equipped throughout in the most practical and thorough manner, and has been called the finest example of horticultural architecture and construction to be found anywhere. The construction is iron throughout. The staging is all of slate and iron; the ventilating apparatus is of the most approved type; the glass the very best; the walks throughout the house are cement. The heating is by means of steam, with boilers fitted to burn both coal and gas. The foundations are fine cut Amherst sand stone. The office building is of the same stone finely executed. Its interior is fitted with offices, work rooms, etc., and is finely finished. Underneath this building is a storeroom, coal storage space, and the steam boilers. Water is delivered at every faucet, both cold and hot; the ponds for lilies are beated with jets of hot water from the water supply system. In fact nothing is omitted to make a model range of glass.

This building, as well as the Conservatory in Allegheny, have been planned by the architects Lord & Burnham Co., of Irving-on-Hudson, and they were executed under their supervision.

The Superintendent of the Pittsburg Public Parks, *Mr. A. M. Bennett*, is also Superintendent of this Conservatory. Mr. Bennett came here from Brooklyn, where he was Superintendent of Melrose Park; he has also occupied a prominent position in the largest commercial nurseries on Long Island : thus he came to his work here with valuable experience. *Mr. Alf. J. Edmonds*, his foreman, came from the U. S. Nurseries in New Jersey. The other assistants are *Mr. Joseph Spring*, in charge of the Palm house; *Mr. David Howells*, department of Tropical Plants; *Mr. Robt. Lunstrom*, Ferns and exhibition houses; *Mr. John Boyle*, Cactus and miscellaneous; *Mr. Chas. Cook*, Aquatics; *Mr. Wm. Frazer*, Chrysanthemums; *Mr. Wm. Goddard*, engineer; *Mr. Robt. Fulton* and *Mr. J. M. Jones*, general assistants. These gentleman have all been selected on account of their special fitness for their duties and the excellent order as well as the remarkable thrift in all departments, gives testimony to their efficiency.

It proved very fortunate that the Schenley Park Conservatory was in shape for the reception of tropical plants just about the time of the close of the World's Fair, when many choice and rare plants, very difficult to obtain at other times and which had been sent to Chicago for exhibition, could be bought at a reasonable price. In order to take advantage of this opportunity Mr. E. M. Bigelow, who has the success of the Conservatory so much at heart, secured a fund, by subscription, from some liberal citizens with which to purchase a number of these plants. The unique collection of Tree Ferns, worth $5,000 was obtained in this way. Other private citizens made valuable additions. Senator Wm. Flinn donated the large and varied collection known as the Drexel collection and worth over $3,000. Mrs. Carr contributed $1,000 worth of plants. Besides there were other welcome contributions from Mr. J. M. Armstrong, Mrs. Frew, Capt. J. J. Vandegrift, Robt. Craig, Mr. R. Gray and others.

SUPERINTENDENT BENNETT AND HIS ASSISTANTS.

BECKERT'S · SEED · STORE,

ALLEGHENY, PA.

The most Complete and best Equipped Seed House in Western Penn'a.

Headquarters for Lawn, Flower and Garden Seeds, Bulbs for Indoor and Outdoor Culture,

CATALOGUES MAILED ON APPLICATION. GENERAL CATALOGUE ISSUED IN JANUARY. CATALOGUE OF BULBS ISSUED IN OCTOBER, ANNUALLY.

W. C. BECKERT, Seedsman.
19 FEDERAL STREET.

---

* * WE RECOMMEND * *

WALLACE OPTICAL CO.

Manufacturing Opticians,

FOR PERFECT SPECTACLES AND EYE GLASSES. . .

. . . 624 PENN AVENUE.

H. A. BECKER,

MANUFACTURER OF AND DEALER IN

MUSICAL @ INSTRUMENTS,

COMPLETE LINE OF
MUSICAL MERCHANDISE.

Band, Orchestra and Sheet Music. Italian Strings a Specialty.

39 CHESTNUT STREET,   ALLEGHENY, PA.

First-class Repairing on all kinds of Musical Instruments.

PURE · ARTIFICIAL · AND · LAKE · ICE
.. AND ..
DISTILLED AND FILTERED WATER.

Natural Ice Capacity, 200,000 Tons.
Artificial Ice Capacity, 350 Tons Daily.

PRINCIPAL OFFICE, THIRTEENTH AND PIKE STS,
PITTSBURGH, PA.

VERSCHAFFILTIA SPLENDI'DA.
Schenley Park Conservatory. (See page 34.)

EXPLANATION.

The descriptive lists in the following parts contain the names of all the plants which the author has been able, with the help of the gardeners, to discover in the two Conservatories.

The botonical names of the plants are given first; there are always two names necessary to designate a plant botanically, the first, a noun, the "genus" name of the plant; the second, the name of the "species," being either an adjective referring to some property of the plant, or the proper name of some person after whom the plant was named, or of some locality in which it is found.

The popular names, as far as such exist or could be ascertained, are stated after the botanical name. Then follows the interpretation of the latter. The letter [A.] indicates that the plant exists in the Allegheny Conservatory; [s.] refers to Schenley Park Conservatory.

There will be, of course, from time to time, changes in the arrangement of plants in the different departments of the Conservatories. It would, therefore, be of little use to indicate the exact place where the various specimens named in the lists could be found at present. All of the more conspicuous and interesting plants are labelled, and it is the intention of the Superintendents to have all others labeled as fast as time and other pressing work will permit. This will enable the visitor to find whatever information this book offers concerning any specimen in the house. It is only necessary to find the name in the alphabetical index at the end of the book, and then turn to the page referred to.

The Palm Houses.

GROUP OF PALMS.
Latania Borbonica above, Areca lutescens and foliage plants below.
Allegheny Conservatory.

The Palm Houses.

NO part of either Conservatory represents tropical splendor and luxuriance more strikingly than the Palm houses. The Palm house at the Allegheny Conservatory has the advantage of age over its mate, or should we say its rival, at Schenley Park. In its overflowing thriftiness it shows the rich results of careful and competent management; indeed the roof would have been too low and the space too narrow, had the plants been allowed to grow and spread at will; constant trimming and removing is necessary to keep the tropical vegetation within bounds and to prevent it from obstructing the passage ways. At the Schenley Park Conservatory wonders have been worked in the short time since this building has been completed. Artificial hills of calcareous tufa have been built, on the tops and sides of which palms and kindred plants have been installed, and they seem to feel quite at home in their new surroundings. In the little recesses, cracks and grottoes, formed by the tufa, Ferns, Selaginellas, Begonias, Peperomias, and other pretty plants are peeping out, and many other cunning devices help to make the scenery look charming in its details, natural and grand in its totality.

But this is only the start; ten years hence we expect the Palm house to represent the scene of a veritable tropical forest. The crowns of the Date Palms, Cocos and Seaforthias will reach far into the lofty dome of the crystal building; the Latanias, Chamærops, Bananas and Cycas will form a roof of verdure above the serpentining walks and the undergrowth of ferns and foliage plants will make the illusion almost complete.

What rare pleasure to walk under the boughs of those grand and at the same time graceful plants. We might imagine to wander amid the Palm groves of India or through some forest on the Amazon.

It is true, the gaudy butterflies of the torrid zone are missing; the birds in gay plumage, the chattering monkeys would be needed to make the illusion even more complete, but we shall not complain, for on the other hand we have the advantage of enjoying this visit to the tropics without having tormenting insects to contend with, nor venemous serpents to fear, or hostile savages to fight.

Let us take a short, general view of the two palm houses, and then discuss the principal plants raised there,—the palms.

Entering the palm house in

SCHENLEY PARK CONSERVATORY,

We are at once confronted by a fine specimen of a fan palm, Lantania Borbonica, less imposing by its height than by its crown of immense leaves ; behind it we notice the peculiar fronds of the Caryota palm, reaching higher at present than its neighbors, the elegant Arecas. To the right is a clump of Rattan palms, and further on several high cones of dark green foliage, the true Laurel or Sweet Bay.

The hills right and left are crowned by large Date Palms, the one on the right the date palm proper, the other to the left, the Spiny Date Palm.

The wall of the office buildings and work-shops is covered with one of the most graceful and grateful of climbing plants, Solanum Seaforthianum, a relative of the potato, as its blossoms readily suggest.

If we pass on to the right, our attention is attracted at the one hand by a fine specimen of Chamærops humilis, and at the other by a graceful tree fern. The entrance to the right wing is guarded on each side by a large climbing plant Philodendron pertusum, (Monstera deliciosa) interesting on account of its perforated leaves.

Farther on a Screw Pine (Pandanus), supported by ærial roots, resembling clubs, is an object of general curiosity, not less so the Silver Thatch Palm, with its natural bandages of woven fibre.

On each side of the western entrance, opposite the alligator pond, stands a fine specimen of the Abyssinian Banana. What beautiful scenic pictures they represent, standing on grotesquely formed banks of tufa, from the niches of which pretty ferns and foliage plants are peeping.

Right near we have the "Monkey Puzzle," Araucaria imbricata, with its symmetrical branches and stout triangular leaves, which must make the climbing of this pine indeed a difficult task.

Proceeding on our way, we admire the handsome Macrozamia, and walking between Fan palms, Cycads and Cabbage palms, we notice the odd Cocoloba, with its immense coppery leaves ; on the right hand the "Little Coco," as the Theophrasta latifolia is sometimes called, deserves notice; it looks indeed like a small palm with its crown of large leaves on its slender trunk.

Passing around the south-west side of the hill, a beautiful picture is presented by the Hanging Fern of the Himalayas, (Goniophlebium sub auriculatum), with its surrounding vegetation; among which the Fan palm to the right of the walk, Sabal mauritiæformis, is of special interest on account of its rarity. The unfolding of the two-ranked crown of leaves of the Traveler's tree in the opposite corner, will be watched with much interest. This tree had fared so badly on its journey from Madagascar, that at its arrival it had only a single leaf left. At the end of last March a second leaf unrolled itself, and others will soon follow.

There are many other beautiful palms and interesting plants thriving finely in the palm house; they will all be mentioned later on in systematic arrangement.

In the
ALLEGHENY CONSERVATORY

We are also greeted at our entrance by a picturesque tropical scene. An immense climbing plant, Monstera deliciosa, with a multitude of ærial roots and rich clusters of fruit, occupies the foreground; behind it we imagine to look into a Banana grove, and indeed, many a bunch of delicious bananas has been raised there. Right and left are finely developed specimens of the Screw Pine, (Pandanus utilis,) and one of the grandest specimens of Cycas Circinalis can be seen in the left corner. The southern wall is densely clad with Clerodendron Balfourii, one of the handsomest climbing plants, quite gorgeous when unfolding its wealth of white and red blossoms.

Following the path, we wander under the verdure of palms of many kinds, among which we find immense Fan palms and a majestic Seaforthia, the crown of which reaches to the glass roof of the building; there are also Cocoanut, Cabbage palms and Date palms, intermingled with Carludovicas, Dragon trees, (Dracæna), Araucarias, Bamboo and Rubber trees, among the latter the handsome, larged leafed Imperial Rubber tree, Ficus imperialis, should not be overlooked.

The rarest specimen in this palm house is Zamia latifrons, which Mr. Hamilton discovered in the Horticultural Hall at the World's Fair. It is said to be the only plant of this kind in cultivation.

Among the foliage plants we notice fine Marantas and Dieffenbachias; but Curculigo recurvata, with its plaited leaf, seems to grow especially thrifty, and forms dense masses of rich verdure.

Cycas circinalis (with 75 fronds); in front at the left, Pandanus utilis; the wall at the left side is covered with Clerodendron Balfouri.
Allegheny Conservatory.

The Palms.

Linnæus, the great botanist, calls the palms the Princes of the vegetable world, and Humboldt speaks of them as the tallest and noblest of all plant forms; the shafts of some specimens he measured himself, proved to be 180 feet in length. Their column-like unbranched stem, which often bears at its summit a crown of gigantic leaves, gives to these plants an aspect of graceful grandeur, and they lend to the scenery in which they naturally grow, a peculiar solemn charm. The palm is the tree of the tropics, and flourishes best where the annual average temperature dwells between 75-82 degrees F. Sóme species, however, reach within the southern limits of the temperate zone.

The flowers of most palms are rather inconspicuous and grow on a compound cluster or spadix, the branches of which resemble catkins; they can often be seen on the specimens in the two palm houses. Besides, most palms are unisexual (diœceous,) having the male or staminate, and the female or pistillate flowers on different trees. They depend for their fertilization upon the wind; insects may also assist. The pollen, of which some palms produce great masses, must often be carried to great distances to reach the pistils of the female plants. The fruit is generally one-seeded, rarely two-seeded; otherwise very variable in size and appearance, from the large cocoanut to the small berry-like fruit of many smaller palms.

The uses of palms are innumerable. In some countries these plants supply nearly all the wants of the natives, and they furnish, besides, many useful articles to the other parts of the world. We need only mention palm oil, wax, sugar, sago, soap, candles, articles of furniture and clothing and food; millions of pounds of palm oil are produced annually in Africa, principally from Elais guineensis; and millions of pounds of palm cane are worked annually into carriages, chair bottoms, brooms and other articles; not less is the consumption of palm fibre for floor mattings, cordage, bagging, coarse cloth, etc. In fact there is no other class of plants so universally useful as the palms.

A poem in the Tamil language mentions 801 uses of the Palmyra palm (Borassus flabelliformis) which grows in southern Hindostan.

There are nearly 1,200 species of palms known, many of these are described in "Seeman's Popular History of Palms."

The palms can be conveniently divided into Fan-Palms or those who have palmately veined, lobed or compound leaves, the principal veins all radiating from one point, and in Feather Palms which have pinnately compound leaves, consisting of a long midrib with usually numerous leaflets growing from its side.

THE FAN PALMS.

Bra'hea glau'ca, (named in honor of Tycho Brahe, the celebrated astronomer; *glauca*, covered with a bloom.) Native of Peru and the Andes. This palm attains a moderate height. Notice the stout spines and the leaf-stalk. [s.]

Chamæ'rops hu'milis, (from *chamai*, dwarf, and *rhops*, a twig, *humilis*, low.) This name has been given to this palm on account of its low growth in comparison with the lofty columns of the tropical palms. This and the cabbage palm are the only species of palm growing native in southern Europe; it does not extend further north than Nice. In Spain and northern Africa it is very common. Its trunk seldom grows higher than three or four feet, and sends up numerous suckers from its creeping roots, thus forming dense thickets which are hard to penetrate, especially on account of the prickly footstalks. Its fruit is about the size and shape of an olive. The leaves are used for making hats, brooms, baskets, also for thatching houses, and from their fibre the French make a material resembling horse hair. The coarse fibre which can be seen at the bases of the leaf-stalk is mixed with camels' hair by the Arabs and woven into tent covers. [A. & S.]

Notice the great amount of fibre around the base of **Chamæ-rops macrocarpa,** (large-fruited.) [s.]

Chamæ'rops staurocantha, (cross-spined), is characterized by the peculiar shape of its spines, which are, however, not as prominent as the spines of many other palms. [s.]

Chamæ'rops excel'sa, (tall), native of Nepaul, is a handsome species, the tallest of this genus. [A.]

Chamæ'rops Fortu'nei, (Fortune's), is a native of China; it is also one of the taller species, and is characterized by its hairy fruit. [A.]

There is a palm of this genus growing in Florida, **Chamæ'rops hystrics,** (bristly), but it is not represented in the conservatories.

CHAMÆROPS HUMILIS.
Schenley Park Conservatory.

Cory'pha austra'lis, (*koryphe*, summit, *australis*, southern.) This is one of the few palms growing in Australia, and is found on the east coast of that continent; it attains a height of over 100 feet, with a trunk about one foot in diameter. The leaves, when still tender, are cooked and eaten; those more mature, but still unexpanded, are scalded, dried and made into hats. (The Coryphæ are among the noblest of the fan-palms, their home is tropical Asia, New Holland and Australia. The Talipot Palm, Corypha umbracolifera (umbrella-bearing), is the most majestic of them all, the leaves, when fully expanded, form a circle of 10-13 feet in diameter, and the leafstalks are 6 to 7 feet long; the trunk reaches a height of 60 to 70 feet and forms a straight, cylindrical column. The leaves are used as umbrellas and tent-covers and large fans made of them are carried before people of rank among the natives of Ceylon. [S.]

Lata'nia Borbon'ica. The Bourbon palm. (*Latania* comes from Latanier; the native name, *Borbonica*, from the island of Bourbon, the native place of this palm. A very fine and popular fan palm. Both conservatories possess beautiful specimens of this species. In Schenley Park we also find

Latania Loddige'sii, Loddiges' Latania.

Latania ru'bra, the red Latania.

Latania au'rea, the golden Latania, from the island of Rodriguez, and

Latania glaucophyl'la, (leaves covered with a bloom.) [s.]

Licua'la hor'rida. (*Licuala*, the native name. From India. If we notice the stout and sharp spines, covering stem and leaf-stalk of this palm, we do not wonder that it is called horrid. [s.]

Not much less forbidding, appears

Licuala spino'sa, (Spiny.) [s.]

and **Licuala acuti'fida,** (acute leaved), a native of the island Pulo-Penang, in the Indian Archipelago. The stems of these palms grow about 5 feet high and 1 inch thick, except at the base, where they are much thicker. They are made into walking sticks, known in England under the name of "Penang Lawyers.'" [s.]

Livisto'nia Chinen'sis, (named after P. Murray, of Livingston, near Edinburg. *Chinensis*, from China.) The Livistonias are allied to the Coryphæ, and are found in southern China, the Indian Archipelago and Australia. Their leaves are divided into numerous segments, which are split at the apex and frequently have threads hanging between them, while the foot-stalks are sheathed at the base in a mass of threaded fibres and are often prickly along the edges. [s.]

Livistonia rotundifol'ia, (round leaved), from the Indian Archipelago, and [A. & s.]

Livistonia Hoogendor'fii are also represented in the Schenley Park Palm house. [s.]

Pritchar'dia macrocar'pa, (named in honor of W. T. Pritchard, the author of "Polynesian Reminiscences." *Macrocarpa*, large fruited.) An ornamental palm from the Sandwich islands, where also,

Pritchardia Gaudichau'dii comes from. [s.]

Pritchardia Pacifica, and

Pritchardia gran dis come from the Polynesian islands. [s.]

Rha'pis flabellifor'mis, (*Rhapis*, a needle; *flabelliformis*, fan shaped, on account of the sharp pointed, fan-shaped leaves.) These palms are related to Chamærops; they are small, with reed-like stems, which grow in tufts from the same roots; they are sometimes called Ground Rattan Palms, and are among the palms which furnish rattan. They are natives of southern China, but are also extensively cultivated in Japan. [A. & S.]

Rhapis hu'milis, (low,) is another species of Rattan palms. Notice that the leaves on the above named palms do not grow in a crown at the top, but alternate and spiral around the stem, and that their sheaths, modified into fibrous tissue, surround the stems as with a network. [S.]

Rhapis Linde'nii is another species found in the Allegheny Conservatory.

Sa'bal Blackburnea'na. Blackburn's Sabal Palm. *Sabal* is probably the native name of the palm. The Sabal palms are next to Chamærops, the most northern of the palms, and all are natives of America. Sabal palmetto, (small palm,) is well known in the South, and its trunks were used for making stockades in the War of Independence. The Palmetto was on that account placed in the arms of North Carolina. [A. & S.]

Thrin'ax e'legans. (*Thrina*, a fan; *elegans*, elegant.) A West India Palm; the specimen in the Schenley Park coming from Trinidad. These are graceful, slender palms, attaining a height of 20 feet. [S.]

Thrinax argente'a. (Silvery.) From Jamaica. Is known as the Silver Thatch Palm; its leaves are not only found excellent for thatching, but they are extensively used for making palm-chip hats, baskets and other articles. The young leaves are eaten as a vegetable. In Panama the palm is called Broom Palm. Visitors to SCHENLEY PARK CONSERVAVORY should notice the beautiful network of fibres at the bottom of the leaves; these fibres represent the sheathing bases of the leaves.

AMONG THE PALMS.
In the centre, Areca lutescens; to the left of it, Caryota urens; to the right below, Latania Borbonica; right and left corner, Laurus nobilis.
Schenley Park Conservatory.

FEATHER PALMS.

Acanthophœ'nix crini'ta. (*Akantha*, a spine ; *phœnix*, date palm ; *crinita*, hairy.) A small palm, with thorny stems and leaves ; its home are the Seychelle and Mascaren islands. This genus is closely allied to the Arecas. [s.]

Arec'a lutes'cens, (*Arec*, the Malay name of the palm, *lutescens*, yellowish.) These palms are known under the name of Cabbage Palms, because the young leaves of some species (Areca oleracea) form a dense head, and while tender, furnish an excellent vegetable. Areca lutescens grows wild in the southern part of France, it is exceedingly graceful and easily cultivated and is therefore one of the most common palms of our green houses, doing good service at every decorative display of plants. [A. & S.]

Areca ru'bra, (red), from the island of Mauritius. [s.]

Areca Bau'eri, Bauer's Areca, from the Norfolk Island. [s.]

Areca Verschaffel'tii, Veschaffelt's Cabbage Palm, from the island of Rodriguez. [A. & S.]

The most important of the Arecas is Areca Catechu, of the East Indies ; it is the palm yielding the Betel nut, which sliced, sprinkled with quick lime and rolled into a leaf of the Betel pepper, forms the favorite chewing material of the people of India ; it is exceedingly acrid and raises blisters on the tongue, gums and palate of the Caucasian at his first attempt of imitating the natives. It has not yet been introduced among the chewing gum sold in this country.

Aren'ga sacchari'fera. (*Areng*, the native name ; *saccharifer*, sugar-bearing.) Is at home on the islands of the Indian Archipelagos, and on account of the variety of its products is of great value to the natives. The black, horse-hair like fibre which surrounds the leaf-stalks is plaited into ornaments, woven into matting and twisted into cords. The sap obtained by cutting off the flower-spikes affords an abundance of sugar which, fermented, turns into toddy or palm wine ; also an inferior grade of sago is produced from the starchy portions of the stems.

Astrocar'yum ensifolium. (*Astron*, a star ; *karyon*, a nut, so called on account of the appearance of the fruit ; *ensifolium*, sword-shaped leaf.) These palms are allied to the cocoanut palm they are natives of South America; their interrupted pinnate foliage is peculiar and their thorny stems make them hard to climb. [s.]

Astrocar'yum Muru-Mur'u, is a Brazilian species, and the graceful

Astrocar'yum australis, grows in Paraguay. [s.]

Atta'lea Cohune. (*Attalus*, magnificent.) The Attaleas are among the loftiest of the South American palms. The species above named is found in Honduras, and produces the Cahoun nuts which yield a valuable oil. [s.]

Attalea excel'sa, (tall), from Brazil; a very young specimen of this species is present.

Calamus ciliaris. (*Kalom*, the Arabic word for a reed; *ciliaris*, fringed.) The Calamus palms, of which there are many species, are nearly all natives of Asia; they have reed-like stems, seldom more than an inch or two in thickness, but of great length, climbing over and amongst the branches of trees and supporting themselves by hooked spines attached to their leaf-stalks; others form low bushes. From different species of Calamus the rattans or canes, so much used for chairs and chair-bottoms, carriages and other purposes, are obtained.

Cayo'ta u'rens. (*Karyon*, a nut; *urens*, stinging. A palm of peculiar aspect, on account of its twice compound leaves, the leaflets of which are triangular to rhomboid, (somewhat diamond-shaped) in outline. In Western Asia, its home, it is a noble and most useful tree, supplying the natives with several important articles. From its flower-stalks a large quantity of toddy or palm wine is obtained, which, boiled down, forms a palm sugar, which is said to be even more delicious than our maple sugar. Nearly all the sugar used in China is obtained from this palm, the cocoanut palm and a few other palms. From the central part of the tree a very good and nutritious kind of sago is prepared, which formed into bread and gruel constitutes a large part of the food of the natives. The fibre obtained from the leaf-stalks, called Kittul fibre, is very strong and is turned to many uses, while the wooly substance scraped off the leaf-stalks is used for caulking boats. [A. & S.]

Caryo'ta sobolife'ra. (Sucker-bearing,) is another species from the Mollucca islands. [A. & S.]

Cerox'ylum niv'eum. Wax Palm. (*Keros*, wax; *xylon*, wood *niveum*, snowy. The young specimen in the Conservatory gives no idea of the beauty and peculiarity of this palm, which is one of the largest of this tribe growing in the Andes of South America. Humboldt mentions of this palm that he found it growing in great numbers near the limits of perpetual snow, the trunk, which attains a great height, is about one foot in diameter, but swells out in the middle to nearly double that thickness; it is covered with a thick coating of a waxy substance, which is scraped off by the natives and mixed with beeswax or tallow for making candles. One trunk furnishes about 25 pounds of this vegetable wax. [A. &. S]

Chamædo'rea el'egans. (*Chamai*, dwarf; *dorea*, gift; " The Dwarf's Gift," so called because most of the palms are so easily reached. It is a native of Mexico; many other species are found in Central and South America. These are low palms, the stems or which are made into walking canes, while the young, unexpanded flower clusters are used in the Mexican kitchen as a vegetable.

[E. & S.]

Co'cos plumo'sus. Cocoa-nut palm. (*Coco* is the Portuguese word for monkey, in reference to the end of the nut resembling a monkey's face; *plumosa*, feathery) This is a Brazilian species. [S.]

Cocos Romanzof'fia. Romanzoff's cocoanut is represented by fine specimen in the Allegheny Conservatory. Admirably graceful are the long primatified leaves of the cocos, the slender prinnae or which are inserted on both sides of the midrib in a peculiar fashion, these prinnae or leaflets are not stiff like those of other palms, but become soft toward the apex so that they partly hang down, giving the frond the appearance of a gigantic plume. The younger leaves are fastened to the stem of the palm by strong fibers; indeed it looks as if they were tied to it with ropes.

The palm which yields the cocoanut of commerce—

Cocos nucif'era (nut bearing), is a noble and lofty tree. Although found in nearly all tropical countries, its home seems to be southern Asia, where it attains its most majestic form, rearing a trunk up to a height of 60 and 80 feet, and attaining a diameter of 2 feet. It is a sea-loving palm, and gives character to the islands of the Pacific, where it is found bordering the low shores, its stately column slightly inclined toward the water. The flowers of the palm are arranged in branching spikes 5 to 6 feet long, and enclosed in a tough covering or spathe. Each spike produces from 10 to 20 nuts. With the fruit and its varied uses the reader is well acquainted. The wood is imported into Europe under the name of porcupine wood, and is used for making articles of furniture. The fiber is extensively used for cordage and matting. The oil extracted from the nut is used for making soap, stearine and for other purposes. [A. & S.]

Cocos Wendellian'a. Wendell's cocoanut, can be found in small specimens among the tropical plants in the right wing of the Schenley Park Conservatory; also in Allegheny Conservatory.

Dæmono'rops palemba'nicus. (*Dæma*, a cord; *rhops*, a twig, so-called on account of the rope-like, climbing stems; the species name refers to Palemba in Java, the home of the palm.) These palms are allied to Calamus which they much resemble. Dæmonorops draco, also known as Calamus draco, furnishes the red resinous substance known as "Dragon's Blood," which is used for dyeing, coloring varnishes, tooth-powders and dyeing horn to imitate tortoise shell. [S.]

Geo'noma Seeman'nii. (*Geonomos*, skilled in agriculture; probably indicating that it takes a skilled gardener to raise these palms.) The Geonoma palms are natives of tropical America where they form part of the underwood of dense forests. Most of them have reed-like polished stems bearing at the summit a tuft of large leaves which are entire when young but afterwards split and become irregular pinnate. The flexible stems are sometimes made into walking sticks. [S.]

Ken'tia Balmorean'a. Balmore's Kentia, (named in honor of the British Lieut-Colonel Kent.) [S.]

Kentia Forsterian'a. Forster's Kentia. [A. & S.]

Kentia Macar'thuri. Macarthur's Kentia. [S.]

Kentia Wendlandian'a. Wendland's Kentia. [A. & S.]

These palms are natives of New Guinea, the Madagascar Archipelago and New Zealand; they are related to the Arecas. Some species of Kentia grow farther South than any other palm.

Martinez'ia caryotifol'ia. (In honor of Balthazar Martinez, a Spanish naturalist; *caryotifolia*, having leaves resembling that of the Caryota.) A native of tropical America, growing to the height of 20 to 30 feet. The mid-ribs and under sides of the leaves are furnished with sharp spines. [S.]

Oreodo'xa re'gia. (*Oreos*, a mountain; *doxa*, glory; *regia*, royal.) As the name implies; the Royal Mountain Glory. This is a lofty palm inhabiting the mountains. It often reaches a height above 100 feet, though its stem is slender. It is a native of the West Indies. [A.]

Phœnicopho'rum seychella'rum. A pretty palm, native of the Seychelle islands, its large leaves are almost entire and their dark colored stalks are bristly with sharp spines. The name might be translated into "date-bearing" but it is said to mean the "palm carried off," because the first specimen that came to the Kew gardens was stolen. [S.]

Phœ'nix dactili'fera. Date Palm. (*Phœnix* is the Greek name of the tree, *dactilifera*, date bearing.] One of the handsomest and most useful of all the palms. Growing in its native soil in northern Africa or southern Asia, the date palm rears its slender, column-like stem to the height of 60 or 80, even 100 feet, supporting a noble crown of 30 to 50 graceful leaves. These leaves consist of a stout mid-rib, bearing numerous leaflets on each side. Toward the base of the leaf stalk these leaflets assume the nature of sharp spines pointing in different directions and protecting the crown and its precious burden of fruit from the inroads of monkeys and small Arabs. It is the leaf of the date palm which is used in southern Europe to decorate the churches on Palm Sunday, and by the Jews at their Feast of Passover. The flowers of the date palm spring from the axils of the leaves in the form of a spadix; they are diœcious, the pistillate and the staminate flowers growing on different trees. To insure an abundant crop artificial fertilization has to be resorted to, a process which seems to have been practiced in the remotest antiquity. The palm begins to bear in about its thirtieth year, then for about 70 years it produces annually 15 to 20 bunches of fruit, each bunch weighing from 15 to 20 pounds. To many of the inhabitants of the oases and of the borders of the desert the date palm furnishes nearly everything necessary to their subsistence. For 9 months out of 12 it is a never failing food supply; its wood is excellent building material; the leaves are used for thatching; the leaf stalks furnish fuel for the kitchen; sugar is made from the sap, also date wine, which, distilled, furnishes a strong drink known under the name of arrac, or toddy; this name is given, however, to any kind of alcoholic liquor produced from palms. [A. & S.]

Phœnix reclina'ta (reclining.) [A. & S.]
Phœnix rupi'cola. (rock-loving.) and [A. & S.]
Phœnix spino'sa. (spiny) are other species represented in the conservatories, P. rupicola being an especially graceful plant.

Ptychosper'ma Alexandra. (*Ptyche*, a fold or winding, *sperma* a seed; alluding to a peculiarity of the seed.) A palm related to Seaforthia; native of Australia. [S.]

Seaforthia elegans. These are among the finest specimens of palms found in both conservatories; their slender and smooth stems have a bulbous thickening at the base where stout secondary roots are often formed, their crown of large pinnate compound leaves, gives the palm indeed an "elegant" appearance and Lord Seaforth, a patron of botanists, may justly be proud of having the genus named after him. [A. & S.]

Verschaffel'tia splend'ida. (Named in honor of M. A. Verschaffelt, who introduced the first known species.) A very interesting small palm which no visitor should fail to examine. It comes from the Seychelle islands, has a dark stem bristling with sharp thorns and supported by a number of aerial roots. The leaf is entire, broad and plaited. The palm stands at present in the wing to the right-hand of the Palm house.

Trachycar'pus excel'sus. *Trachys*, hairy; *Karpus*, fruit; so-called on account of its rough, hairy fruit.) This is the most interesting of the Fan palms in Allegheny Conservatory; it is a late arrival and therefore added here out of place. The specimen was exhibited at Chicago by the University of Tokio in Japan; it consists of a dozen palms grown upon an old fern trunk and represents an odd and rare curiosity.

Other Large Plants in the Palm Houses.

SCREW PINES.

These plants, which have a somewhat palm-like appearance, are interesting on account of their peculiar leaf arrangement. The long, rigid leaves, which are armed along their edge and midrib with closely set and sharp prickles, are arranged in three ranks, each rank forming a spiral along which the leaves are placed in close succession, thus forming a threefold screw. Another habit worth noticing is that of producing large aerial roots which aid in supporting the plant as it grows in size. The Pandanaceæ occupy botanically a position between the aroids and the palms. To call them "pines" is a misnomer and misleading in regard to the relationship of the plant, as so many popular names are. The pineapple for instance is neither the fruit of a pine nor that of an apple. The fact that the fruit is edible and that it somewhat resembles a large pine-cone gave rise to the name; the Pandanus has a fruit somewhat resembling the pine-apple but more globular; to distinguish it from the latter, which grows in the same localities, it was called screw pine. Most species of this interesting family are found in the islands of the Indian Archipelago where they cover large tracts of country with an almost impenetrable mass of vegetation.

Panda'nus ut'ilis. (Screw Pine.) *Pandang*, the Malay name; *utilis*, useful.) The home of this species, of which handsome specimens are found in both conservatories, is the island of Mauritius; it is the most useful of the Pandanæ, the fiber obtained from the leaves forming excellent material for making strong sacks for exporting sugar and bags for the shipping of coffee. [A. & S.]

Pandanus Veit'schei. Veitch's Screw Pine, is a native of Polynesia. [A. & S.]

Pandanus Java'nica variegata. Variegated Javanese screw pine. A smaller species with white margins to the leaves. [S.]

Pandanus reflexus, (bent back leaves) from the East Indies. [A.]

Pandanus graminæfol'ia, Grass-leaved Screw Pine. A handsome form with narrow leaves. [A.]

Related to the Screw Pines are the Carludovicas, favorite ornamental plants with large plaited leaves, somewhat resembling those of the fan palm: the flowers are borne on a club-shaped spadix, supported on a straight scape. This plant furnishes the material for the Panama hats.

PANDANUS UTILIS.
Allegheny Conservatory.

Carlud'ovica palmata. (Named after Charles IV. of Spain and Louisa, his queen. *Palmata*, hand-shaped leaved.) This is the best known and also the handsomest of this genus; there is scarcely a conservatory or a greenhouse of any pretentions which has not some specimens of it. It is a native of Peru. [A. & S.]

Carludovica rotundifol'ia, (round-leaved) is represented in the Allegheny Conservatory also

Carludovica atrovir'ens, (dark green), while the Schenley Park Conservatory possesses a specimen of

Carludovica hum'ilis, (low) from New Granada.

THE BANANA.

Although stately and palm-like, the bananas are in reality only overgrown herbs. They have a false trunk, formed by the combined sheathing bases of the leaf-stalks. The material of this trunk or shaft is soft and fibrous, so that the thickest "banana tree" can be felled with a table-knife. In the banana plantations the stems are cut away near the ground after the plant has matured its fruit. New shoots appear at the sides of the old stem; these attain their full growth in from six to nine months, yielding another abundant crop of their excellent fruit.

According to Humboldt, the productiveness of the banana as compared with wheat is 133 to 1, and as against potatoes, 44 to 1. Taking equal weights of potatoes and bananas, the latter are about twice as nutritious.

The banana seems to have come originally from Asia; whether it grew indigenous in America or has been introduced at an early date is still a matter of dispute. At present the bananas are cultivated in all tropical climates where fertile soil and moisture are abundant; they form an important and wholesome food supply for many of the inhabitants of the tropics, especially in the islands of the Pacific. From Stanley's reports and experiences we know what a blessing this plant is for the inhabitants of Central Africa. The flowers of the banana grow on pendent spikes; at first they are protected by large, highly colored bracts; as these bracts drop off one by one, they disclose groups or rows of tubular flowers, from 5 to 12 or more in number. As the ovaries develop to form the fruit, they grow upward, a peculiarity not shared by many kinds of fruit.

Not only the ripe fruit of the banana is eaten; the unripe fruit, which consists almost entirely of starch, is used in immense quantities to prepare banana or plantain flour. For that purpose the unripe fruit is dried in the sun and ground in mortars. The young shoots of the plant are eaten as a vegetable and the fermented sap is said to afford a pleasant beverage.

The adoption of the banana in this country as a favorite "all-the-year-round" fruit dates only a few decades back, and in Europe it is as yet scarcely known except at the seaports. We get our supply chiefly from the islands of the West Indies.

The varieties of the edible banana also called plantain, are numerous, but only two species have been thus far distinguished: Musa sapientum, the banana which comes into our markets, and Musa paradisi'aca, the plantain or "Fig of Paradise." This name has its origin from a notion that the plantain was the forbidden fruit which brought our ancient relatives into trouble.

Mus'a sapien'tum, Banana. (*Musa*, from Mauz the Egyptian name; *sapientum*, wise man. It is distinguished from Musa paradisaica by shorter and rounder fruit and purple spots upon the stem. [A. & S.]

Musa tex'tilis. Manilla plantain. This is the plant from which the Manilla Hemp is prepared as well as Manilla paper and other useful material. [A. & S.]

Musa Cavendishii. Cavendish's Banana. A dwarf species from China, very largely grown in conservatories and gardens on account of its fine foliage. [A. & S.]

ABYSSINIAN BANANA.—Schenley Park Conservatory.

Musa enset'e. Abysinian Banana. A handsome species; its fruit is not edible, but the base of the flower stalk is cooked and eaten by the natives. [S.]

Musa vittata variegata. Striped and blotched with white. [A.]
Some of the plants related to the bananas are:

Ravena'la Madegascarien'sis. The Traveler's Tree. A beautiful palm-like plant with immense oblong leaves which spread out like the vanes of a gigantic fan; it is a native of Madagascar. A considerable quantity of water is stored up in the large, cup-like sheaths of the leaf-stalk, a provision most welcome to the exhausted and thirsty traveler and explorer in the wilds of Madagascar. A visitor to that island, Rev. William Ellis, says he was skeptical about the stories told of the Ravenala but on one of his expeditions they came to a clump of these trees; one of his bearers thrust his spearhead several inches deep into the thick, firm end of the leaf-stalk, where it joined the trunk. Instantly a stream of water gushed forth, about a quart of which was caught in a pitcher. The water was cool and perfectly sweet. The fruit of the Traveler's tree is edible, the leaves form excellent thatch for the huts of the natives and the stems are used for partitions and flooring. [s.]

BIRD OF PARADISE PLANT.
Schenley Park Conservatory.

Strelit'zia regin'æ. Bird of Paradise Plant. (*Reginæ*, the queen; named after Charlotte, queen to George III. of the house of Mecklenburg-Strelitz.) This interesting plant is a native of the Cape of Good

Hope. To appreciate its beauty one has to see it in flower. The large bud is placed obliquely on a long flower-stalk which may be compared with the slender neck of a bird, while the head resembles a bird's head, and the perianth, consisting of orange-colored sepals and purple petals, imitates the crown-feathers of the fancied bird. The seeds of this plant are eaten by the Kaffirs. [A. & S.]

Strelitzia augusta, (grand) is another fine representative of this genus, with handsome, large leaves which remind one of the foliage of Ravenala. [A.]

THE CONIFERS.

Some of the tropical Conifers, the Cycads and Zamias, have much resemblance with palms, also with tree-ferns while others, the Araucarias, resemble more our pines, but have generally stouter and broader leaves.

Cyc'as circinal'is. (*Cycas*, Greek name for a palm; *circinalis*, the young leaves rolled up like the frond of a fern.) Native of the East Indies. See page 22. [A. & S.]

Cycas revolut'a. (rolled back-leaved) Native of China. [A. & S.]

Cycas Nove Caledon'ica. From New Caledonia. [S.]

The Cycas have an ancient pedigree; they appeared at the close of the carboniferous age, and reached their greatest development during the age of reptiles. Botanically the Cycas seem to form a link between the the exogenous and the endogenous plants. They are of slow growth but long lived; they cover large areas in southern Africa, where they furnish the "Kaffir bread." The largest and handsomest species are found in the Moluccas, but they have representatives in nearly all parts of the tropics. Three Cycas occur in Florida: Zamia Integrifolia, Zamia pumila and Zamia Floridana. Cycas revoluta and some other species contain considerable quantities of starchy matter in their thick stems, from which a kind of sago is prepared; therefore they are sometimes wrongly called Sago palms; the real Sago palm is Sagus Rumphii.

Zamia integrifol'ia (entire-leaved) is a dwarf species growing in the West Indies, but it is in southern Africa where this genus abounds so as to form a conspicuous feature in the vegetation of that country. [A. & S.]

Macrozam'ia lat'ifrons. (*Macros*, long, great; *latifrons*, broad-leaved.) This interesting plant has been mentioned before. Mr. Hamilton, Superintendent of the Allegheny Parks, noticed it among the horticultural exhibits in Chicago and recognized it at once as a species not heretofore described. Mr. Nicholson, director of the Kew gardens near London, one of the first authorities in botanical matters, gave it the above name on the account of its broad leaves. [A.]

Macrozamia furfurac'ea. Schenley Park Conservatory possesses probably the finest specimen of this most handsome Australian species. (See Illustration.)

Macroza'mia spiral'is. (*Spiralis*, spiral-leaved.) The Macrozamias are nearly all natives of Australia, the midrib of the young leaves of the above species is twisted, forming a spiral. [S.]

Araucar'ia imbricat'a. Chili pine, or Monkey puzzle. (*Araucaros*, the name given to the tree in Chili; *imbricata*, overlapping, like shingles.) These beautiful trees form vast forests in the mountains of southern Chili; they attain great height and are valuable for timber. The seeds are edible when fresh and grow in large cones. It is quite hardy and fine specimens adorn many a park in England and Ireland. It does not seem to stand our climate. On account of its rigid pointed leaves it is even for monkeys a hard task to climb it. [S.]

Araucaria Bidwil'lii is another noble tree of wonderful symmetry, it also grows in South America and bears immense cones the seeds of which are eaten by the natives.

Araucaria Cunningham'ii, from Moreton Bay, is quite different in appearance from the other species and has a white bark; a fine specimen can be seen in the center of the Fern house in Schenley Park Conservatory.

Ced'rus deodar'a. Deodar or Indian Cedar. The name Cedrus seems to be derived either from the Arabic *Kedron*, power, in reference to its majestic appearance or from *Cedron*, a brook in Judea. The above named species is a native of Nepaul and is often propagated for parks by grafting it on the common cedar. [A.]

Macrozamia furfuracea, Schenley Park Conservatory.

VARIOUS OTHER SPECIES CULTIVATED IN THE PALM HOUSES.

Most of the smaller plants in the Palm Houses are also found, and in greater variety, in the other houses and they will be mentioned later on, but the larger plants, not yet enumerated, are included in the following list.

Agav'e Americana. American Century plant. *(Agavos,* admirable.) The Century Plants are mostly natives of Mexico and South America ; they are characterized by their large fleshy leaves with spiny margins and tips. Some of them are long-lived plants growing slowly until the flower-stalk forms, which rapidly rears its chandelier-like flower-cluster up to considerable height ; from 15 to 20 and even 40 feet. After having expanded its multitude of yellowish flowers which spread fragrance in every direction, and after having ripened its seed, the plant is exhausted and dies. The age which these plants attain depends upon the species and upon the climate and conditions of growth. While in rare cases they may live one hundred years before they bloom, most attain the glorious end of their patient lives in 30 to 60 years or about the average lifetime of a human being. [A. & S.]

Agave glauca. (*glauca,* the leaves covered with a bloom.) This is a peculiar species, much resembling a palm ; it forms a stem 15 to 20 feet high on which there is a crown of numerous narrow leaves while the old dry leaves below gradually break away. Two of this species are at present in the aquatic department of Schenley Park Conservatory ; they come from Mexico. From the fibres or several Agaves ropes and paper is manufactured and Humboldt describes a bridge with a span of 130 feet over the Quimbo at Quito of which the main rope, four inches in diameter, was made of Agave fibre. From the concentrated juice of the leaf a kind of soap is made which will lather in salt water as well as in fresh water. But one of the most important products from the Mexican point of view is the favorite beverage Pulque, which is prepared by cutting out the inner leaves just before the flower-stalk is bursting out ; a considerable quantity of sap flows out at the wounded spot, which is of slightly acid taste and easily ferments, assuming a very disagreeable odor while the taste somewhat resembles cider and is pronounced delicious by those who are used to it. The name Agave and Aloe are often confounded because the plants have some similarity. Agaves belong to the Amaryllis family, while Aloes belong to the Lily family.

Arun'do don'ax variegat'a. (*arundo*, a reed; *variegata*, variegated.) this is the tallest grass growing in Europe; in Spain and southern Italy it reaches a height of from 15 to 20 feet; some variegated forms resemble a gigantic ribbon grass and are very handsome. The hollow stems are made into flutes, pipes and fishing rods. It is said that the heroes of Homer made arrows from this reed and that Achilles thatched his house with its leaves. A fine and tall specimen of this plant is in the Palm House in Allegheny Conservatory; those in the Schenley Park Conservatory are at this writing not yet fully developed.

Bambus'a arundinac'ea. Bamboo. (*Bamboo*, the Indian name; *arundinacea*, reed-like.) The tallest and one of the most useful of all grasses. In Eastern Asia, its principal home, it attains often a height of 50 to 60 feet in one season. The Chinese and Japanese build their homes with Bamboo canes, manufacture their furniture from it, equip their ships with masts and beams and sails made from this material and use it for innumerable other purposes for which its lightness and strength adapt it. Bamboo cane finds also in our country various uses among which that for fishing rods is not the least important. [A. & S.]

Beaucar'nea recurvat'a. (*Recurvata*, turned back; on account of the leaves being turned backwards.) A handsome, yucca-like plant from Mexico; the base of the stem is bulb-like; the flower-cluster which forms at the top of the leafy crown is sometimes over a yard in height and bears from 4,000 to 5,000 small fragrant flowers. It is a member of the Lily family. [A. & S.]

Dasylir'ion glaucophyl'lum. (*Dasys*, thick; *lirion*, a lily.) This plant resembles and is related to Beaucarnea and has also, as the name implies, light green leaves, covered with a whitish bloom. It is also a native of Mexico. [S.]

Coccol'oba pubes'cens. Sea-side grape. (*Kokkos*, a berry; *lobas*, a lobe, in reference to the fruit; *pubescens*, downy.) A handsome plant with large, sessile leaves which turn to a copper red; it is situated under the large Date palm on the hill to the left in Schenley Park Palm House. The plant more properly known under the name of "Seaside Grape" is Coccoloba uvifera (grape-bearing) in which the outside flower parts—perianth, become pulpy and of a violet color and surrounds the fruit; it is edible and of a pleasant acid flavor. It is a native of the West Indies. Knotweed family. [s.]

Al'oe Socotrin'a. (*Alloeh*, the Arabic name of the plant; *socotra*, from the island of Socotra.) The aloes belong to the Lily family and are found in nearly all tropical countries but especially in southern Africa. They are largely cultivated for the drug prepared from the dried bitter juice of the plant and known under the name of Aloes. The species named above is said to yield the best quality of that drug. [s.]

Cupan'ia filicifol'ia. (Named after F. F. Cupani, an Italian monk, who wrote on botany; *filicifolia*, fern leaved.) A small tree which by many visitors is taken for a tree fern because its compound leaves resemble the fronds of ferns. The Cupanias belong to the Soapberry family and many of its species are useful trees yielding valuable woods such as the "Tulip wood," the "Loblolly wood" and others. [A.]

Dracaen'a drac'o. Fine specimens of this species are found especially in the Allegheny Conservatory. Dracaenas are all mentioned further on under "Foliage Plants."

Eucalyp'tus glob'ulus. Gum Tree. (*Eu*, good; *Kalypto*, covering, referring to the peculiar calyx which covers the flowers when in the bud; *globulus*, round, globular.) The Eucalyptus trees are natives of Australia and Tasmania and are of great importance to these countries furnishing not only the principal and very valuable timber, but also an aromatic gum and a number of valuable medicinal products. One species, Eucalyptus Ganni, is called the Tasmania Cider tree, as it yields a cool, refreshing liquid from cuts made in its bark during spring. Eucalyptus gigantea is the giant tree which attains a height of 400 feet and above with a circumference of 100 feet near the ground. Eucalyptus trees are cultivated now in great numbers in California and in Southern Africa; they are said to exert a healthful influence upon the atmosphere of the surounding country. Myrtle family. [s.]

Euon'ymus variegat'us. Spindle tree. (*Eu*, good; *onoma*, a name, literally, of good repute.) The spindle trees of which we have several species in this country, known under the name of

"Burning Bush," "Strawberry Bush," "Wahoo," are most beautiful in the fall, when their three to four-celled fruit capsules burst open and show the bright red seed. The above is from Japan. The name Spindle tree is due to the fact that spindles were made from its wood, which is hard and tough and still much used for making shoe pegs, toothpicks and various other articles. [s.]

Fic'us elas'tica. India Rubber Tree. (*Ficus*, fig tree.) Its dark green, thick and glossy leaves, and the bright-colored bracts which envelop the terminal bud make it an ornamental plant. We would scarcely recognize it in its home in India, where it attains the height and thickness of a majestic tree, from which masses of thick aerial roots are hanging down. It is by making incisions in these roots, principally, that the milky sap is obtained which yields the rubber. The native trees are fast becoming destroyed by the reckless treatment of the sap-gatherers; but the regular cultivation of the rubber tree has been commenced. From the age of 40 years a tree yields about 40 pounds of coutchouc a season. It is only safe, however, to tap them once in three years. The India-rubber tree, Ficus elastica, belongs to the order Moraceæ, or the Mulberry family; it has a number of interesting relatives: the common edible fig, Ficus carica, the Mulberry, the Breadfruit tree, the Cow tree or Milk tree of Venezuela, the Osage orange, the Banyan tree, Ficus Indica, the ill-famed Upas tree of Java, and others. [A. & S.]

Ficus elas'tica variegata. [A.]

Ficus imperial'is. A very handsome species with large leaves. [A. & S.]

Ficus Parcel'li. (*Parcell's.*) Leaves variegated with white and green; Polynesian Island. [A. & S.]

Ficus glomerat'a. A narrow leaved species. [s.]

Ficus rep'ens. (Creeping) From China. [A. & S.]

Ficus scan'dens. (Climbing) India. [s.]

These two last species are small and can be seen climbing and trailing like ivy among the rocks in Schenley Park Palm house.

Ficus min'imus, (least) resembles Ficus repens and is also from China. [A.]

Ficus Chauvie'ri, (*Chauvier's Fig tree*) grows to considerable size and can be raised as easily as Ficus elastica.

Gaston'ia palmat'a. (Named in honor of Gaston de Bourbon, son of Henry IV.) It is a native of Mauritius and belongs to the Ivy family. [A.]

Hibis'cus Cooperi. Cooper's Hibiscus. (*Hibiscus* was Virgil's name for the Marsh Mallow.) This is a fine shrub with large scarlet flowers; a native of New Caledonia. The one in Schenley Park is of the variety tricolor. [A. & S.]

Hibiscus rosa-sinensis, (Chinese Rose) is another beautiful specimen from the East Indies. [S.]

The "Marsh Mallow" of this country is Hibiscus Moscheutos, a tall herb with large pink flowers, it grows abundantly in marshy places on the eastern coast of the United States and also near the Great Lakes. The seeds contain a mucilaginous substance from which it is alleged the marsh mallow drops of our confectionary shops are made. Mallow family. [S.]

Justic'ia car'nea. (Named after a celebrated Scotch horticulturist, J. Justice; *carnea*, flesh-colored.) A shrub bearing two-lipped, white and pink flowers; native of India and Southern Africa. Acanthus family. [S.]

Justicia bicalyculata, (having two cups or flower-coverings.) [S.]

Lau'rus nob'ilis. (From the Celtic *blaur* or *laur*, green; *nobilis*, noble.) The genuine Laurel of southern Europe, the leaves of which, arranged in wreaths were used to crown honored poets and heroes. The aromatic oil contained in the leaves make them useful in various other ways. Schenley Park Conservatory contains several very fine specimens of the plant. Laurel family. [A. & S.]

Monstera deliciosa also called

Philodendron pertusum is described later on among the foliage plants.

Pittospor'um Tobir'a. (*Pitte*, pitch, tar; *sporos*, a seed; so named on account of the resinous coating of the seed; *Tobira*, the native name of the plant.) A handsome shrub from Japan, with a profusion of white flowers whose fragrance resembles that of the orange blossom. It is a favorite plant in the Paris flower gardens. There is also a variegated specimen of this species in Schenley Park. Pittosporum family. [S.]

Sansevie'ra Zeylan'ica. Bowstring Hemp. Named after Raymond de Sansgrio, Prince of Sanseviera. *Zeylanica*, from Ceylon.) A peculiar plant with long and stiff, somewhat cactus-like leaves which are cross-striped with dark green and whitish bands.

The fibres of these leaves are used for bowstrings by the natives of Ceylon and other countries where the plant grows. The African Bowstring Hemp is Sanseviera Guineensis. [A. & S.]

Sanseviera Japonica. From Japan. [S.]

These plants belong to the Lily family.

Theophras'ta latifolia (Named after Theophrastus Paracelcus; *latifolia*, broad-leaved). This small tree with slender trunk and crown of rich, broad leaves has already been referred to in the introductory remarks on the Palm houses. It is a native of South America. Myrsina family. [A. & S.]

Toxicoph'laea Thurnber'gii (*Toxico*, poison; *phleros*, bark). A fine shrub, but very poisonous; a decoction of the bark was used by the Bushmen of South Africa for poisoning their arrows. It is a native of the Cape of Good Hope. [S.]

Toxicophlaea spectab'ilis (showy) is from Natal. Dogbane family. [A.]

Nerium olean'der, Oleander. (*Neros*, moist, referring to the nature of the place in which it grows). The Oleander has long been popular as a garden shrub on account of its showy appearance when in bloom. It belongs to the same family as Toxicophlaea and is also very poisonous. [A. & S.]

DEMMLER BROTHERS,
526-528 SMITHFIELD ST.
PITTSBURGH, PA.

House Furnishing Goods

AND..
..THE..
..BEST

NOVELTIES
FOR THE...
KITCHEN.

AGENTS FOR THE CELEBRATED

·· ALASKA AND NORTH STAR ··

·· REFRIGERATORS ··

PASTEUR AND ALLEN WATER FILTERS...

.. The Great PERFECTION AND JEWEL GAS RANGES

GASOLINE, OIL AND GAS

COOKING STOVES.

GAS AND OIL

HEATING STOVES.

Water Coolers, Ice Cream Freezers, Bird Cages, Washing Machines, Wringers, Carpet Sweepers, Garbage Cans and Pails, Brass Goods,

·· AND AN ENDLESS VARIETY

· OF OTHER USEFUL ARTICLES.

JAMESTOWN ART ASSOCIATION,

ROOMS 28 TO 32 McCANCE BLOCK,

Cor. Smithfield St. and 7th Ave.

TAKE ELEVATOR. PITTSBURGH, PA.

PASTELS.

SEPIAS--PORTRAITS—OILS.

WATER COLORS.

FREE HAND CRAYONS A SPECIALTY.

FRAMES WHOLESALE AND RETAIL.

YOU ARE CORDIALLY INVITED TO CALL W. M. STON
..AND..   AND   } Mgrs.
INSPECT OUR WORK.   J. G. SMITH,

ESTABLISHED 1842.

TAYLOR & DEAN,

MANUFACTURERS OF

ORNAMENTAL

IRON AND WIRE WORK

OF EVERY DESCRIPTION.

OFFICE AND WORKS,

201 to 205 Market Street,

PITTSBURGH, PA.

JAMES F. McMORRIS,

Successor to JOS. EINSTEIN & Co.

IMPORTER AND BOTTLER,

NO. 52 SIXTH AVENUE,

ONE DOOR FROM SMITHFIELD ST.   PITTSBURG, PA.

SPECIALTIES.

Berghoff Brewing Co's Dortmunder and Salvator Beers.

Bartholomay Brewing Co's "Bohemian" Beer.

Anheuser-Busch "Faust" Fresh, Bottling and Export Beer.

Eureka Mineral Spring Co's Natural Mineral Water and Ginger Ale.

Gray Mineral Water, Cambridgeboro, Pa.

Imported Kaiser-Culmbach and Pilsner Beer.

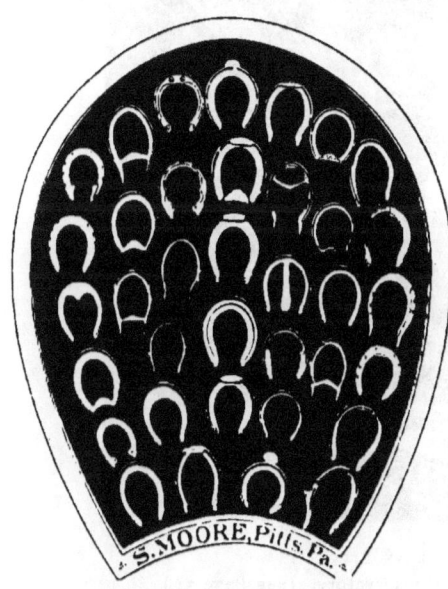

S. MOORE,

HORSE SHOER,

5813-5815 Penn Ave.

East End,   Pittsburg, Pa.

TELEPHONE B. B. 237.

Also Manufacturer of the

Eureka Hoof Ointment and Liniment.

Eureka Healing Salve.

Monitor Rheumatic Liniment.

GERMAN, .   HIGHEST
     FRENCH, .   RECOMMENDATIONS.
          ENGLISH, .   NATIVE TEACHERS.
BEST INSTRUCTION   SPANISH, .
CONVERSATION.   ITALIAN, .
ELOCUTION,   LATIN, .
                                        GREEK.
        . BERLITZ METHOD .
   SCHOOL OF LANGUAGES.
           PENN BUILDING.
     OFFICE AND BUREAU OF TRANSLATION, ROOM 503.
     DIRECTOR. PROF. RICHARD A. SKALWEIT.
SEND FOR CIRCULAR.

PHŒNIX SPINOSA,
Schenley Park Conservatory. (See Page 34.)

Foliage Plants.

Dracaena terminalis. Marantha zebrina. Sphaerogyne latifolia. Croton Challenger. Strobilanthus Dyeriana.
Sanchizia nobilis. Dieffenbachia Bausii. Dracaena amiabilis. Calladium marmoratum. Marantha regalis. Dracaena Lindeni.
Croton Andracanum. Anthurium crystallinum. Dracaena Hybrida.

GROUP OF FOLIAGE PLANTS.   SCHENLEY PARK CONSERVATORY.

Foliage Plants.

In this chapter are collected the plants which are cultivated rather for their attractive and ornamental foliage than for their flowers; some of them, as the Cannas, Anthuriums and others combine a showy flower with a decorative leaf. In the conservatories some of the plants named in the following list are raised in the Palm houses while the majority can be found in the other tropical departments. We shall begin with the plant belonging to the

ARUM FAMILY.

Because they constitute a great part of the foliage plants and are characterized by an odd manner of bearing their flowers and fruit, somewhat resembling the inflorescence of the Palms and Bananas. The flowers are borne upon a fleshy club called a spadix and generally surrounded by an envelope, called a spathe. The Calla, wrongly called Lily and the Jack-in-the-Pulpit of our woods are familiar examples.

THE DIEFFENBACHIAS.

The plants of this genus, which is named after the German botanist Dr. Dieffenbach, are characterized by handsome, often variegated foliage. But they possess an exceedingly acrid and poisonous juice and if part of the plant should ever be brought into the mouth, it causes the tongue to swell and the mouth to burn so intensely that speech is impossible. Many fine specimens of this are found in the conservatories; they represent the following species:

Dieffenbach'ia Baraquinian'a (Baraquin's), from Brazil. The leaf has a white leaf-stalk and midrib. [A. & S.]

D. Baus'ei (Bause's). The leaf with its dark green margin and yellowish green center sprinkled with white spots is a beautiful object. [A.]

D. Bowman'ii (Bowman's). From Brazil. The leaves are yellowish green blotched with dark green. [A.]

D. Imperial'is (Imperial). From South America. The leaves are dark green with yellow spots, midrib grayish. [S.]

D. Memoria-Cor'si has a green leaf with yellow lines and dots starting from the midrib.

D. nob'ilis (noble). From Brazil. Leaves with central grey band and yellowish green patches. [A. & S.]

D. pic'ta (painted). Tropical America. Leaves white spotted. [S.]

D. Weir'ii (Weir's). From Brazil. Leaves marbled with yellow. [S.]

D. Seguin'e (Seguine's) not represented in the conservatories; has been used by the slaveholders to punish their slaves; they were forced to bite the plant, which caused agonizing pain and deprived them for some time of the power of speech; hence the name "Dumb cane."

THE ANTHURIUMS.

The name of this plant is derived from two Greek words meaning "Flower-tail;" it is suggested by the long, tail-like spadix which contains numerous perfect flowers. The spathe of some species is brightly colored. Formerly the plants were called "Pothos" under which name they are described by Humboldt and other travelers in South America. They are found sometimes growing between the forks of trees and sending down a cluster of aerial roots; being air plants like many orchids. In the following list of Anthuriums which can be seen in the Conservatories the species names are not translated since they are nearly all proper names.

Anthurium Andraean'um, has a bright scarlet spathe and a yellow and white spadix. [A. & S.]

A. Augustinum. From tropical America [A. & S.]

A. Brownii. [S.]

A. Clarkian'um. [A.]

A. Crystallin'um has a most handsome leaf of a velvety green color with white veins (shown in the illustration). From Columbia. [A. & S.]

A. Ferrierense has a pink spathe and dark purple spadix.

A. grande. [A.]

A. Lindenian'um, satiny leaves, a white spathe and purplish spadix. [A. & S.]

A. magnif'icum. [S.]
A. Reynoldian'um. [S.]
A. Rotschildian'um. [A. & S.]

A. Scherzerianum, has a scarlet spathe and a twisted spadix and is popularly called "Flamingo plant." It comes from Costa Rica. [A. & S.]

Caladium marmorat'um. (The name, probably, is derived from *Kaladion*, a cup; *marmoratum*, marbled.) Among the foliage plants few are more decorative than the Caladiums; they are just'y popular and are planted in large groups in parks and on private grounds, they present a fine tropical aspect. The root-stalks of these plants are rich in starch, but most of them are very acrid and some even poisonous, but Caladium bicolor, C. sagittifolium are used as food. C. esculentum is even cultivated in abundance in India and brought to the markets; the leaves as well as the root-stalks being used for food. These plants are popularly known under the name of Elephant's Ears.

The Allegheny Conservatory possesses a fine assortment of "Fancy-colored Caladiums" which in gorgeousness and variety of color are scarcely surpassed by the most brilliant flower.

Related to the Caladiums are the Alocasias; these plants also produce edible tubers. They are distinguished by their handsome arrow and shield-shaped leaves on long, stout leaf-stalks. They are natives of India and the islands of the Pacific. The name Alocasia has been chosen to distinguish them from *Colocasia* (the Greek name for the root of an Egyptian plant) to which they are closely allied. Handsome specimens can be seen in the Schenley Park Palm house among which the following:

Alocas'ia arbor'ea (tree-like). [A. & S.]

A. macrorhiz'a, (*macros*, long; *rhiza*, root; having a large root-stalk). From the Polynesian islands. [A. & S.]

A. metal'lica (metallic) The Shield Plant. The leaf is of a rich bronze hue and resembles a metal shield. From Borneo. [S.]

A. Sanderian'a. [S.]

A. Seden'ii (Seden's). Leaves arrow-shaped, bronzy green, purple beneath; veins white. [A. & S.]

A. Veit'chii. [A. & S.]

A. violac'ea (violet). [S.]

A. Walton'i. [A. & S.]

A. zebrin'a (zebra-like). From the Philippine islands. [A.]

Richard'ia Ethiop'ica. Calla Lily. (*Kalos*, beautiful.) This handsome and popular plant is too well known to need description. While commonly called a lily it does not belong to the Lily family, having no calyx composed of petals, but a spadix on which the stamens grow above and the pistils below and which is surrounded by a large white spathe. The plant is named after Richard, the noted French botanist. It is a native of the Cape of Good Hope.

A'corus variegat'a. (*a*, not ; *kore*, the pupil of the eye ; the name refers to some medicinal property of the plant; *variegata*, variegated.) Well known is Acorus calamus, the Sweet Flag, the aromatic root-stalks of which many of the readers may have chewed in their early chiidhood.

Aglaonem'a commutat'um (*Aglaos*, bright; *nema*, thread ; *commutatum*, changed). The leaves are blotched with grayish spots ; the spathes are fragrant. A native of the Philippines. [A.]

Aglaonema pic'tum (painted . Leaves with light and dark green designs. [S.]

Curmer'ia picturat'a (Painted Curmeria). This is a relative of the Richardia, it has on its heart-shaped leaves a broad, central band of silver gray, [S.]

Curmeria Wallisii Wallis') has a dark green leaf with yellowish green.

Monster'a deliciosa (On account of its delicious fruit). Also called

Philoden'dron pertus'um (from *Philo*, to love, and *dendron*, tree). This is a mighty climbing plant, producing a tangle of aerial roots. The specimen facing the visitor of Allegheny Conservatory at the entrance to the Palm house gives an excellent idea of the appearance of the plant in its native place, the primeval woods of Mexico and Southern America. Its fruit resembles a large ear of corn and the pulp which surrounds it, when ripe, has a delicious pineapple-like flavor. The specimen mentioned yields abundant fruit every year. The plants in Schenley Park Conservatory are not quite as large, but are growing rapidly. The peculiarity of the leaf, which is perforated, has been mentioned before.

MONSTERA DELICIOSA AND GROUP OF BANANAS.
Allegheny Conservatory.

Philodendron Sellowian'um (Sellow's) has a white spadix and a greenish white spathe. [A.]

Schismatoglot'tis Laval'lii (*schisma*, falling away; *glotta*, a tongue; because the upper part of the spathe falls off so quickly). Natives of Borneo. A fine foliage plant resembling a Maranta. A splendid specimen of it can be seen in the Palm house. [s.]

THE CANNAS.

The Cannas, natural relatives of the Bananas, share not only the luxuriance of the children of tropics, but lend besides the charm of glowing color. Like torches of red and golden flame they glow in various places in the Palm house and other tropical departments and seem almost to light up their surroundings. The Cannas, sometimes called "Indian Shot" on account of the round, black and hard seeds, have become very popular of late and there is scarcely a garden to-day without some group or groups of this rich, ornamental plant. Much has been done too, especially by French nurserymen, to produce new, handsome and showy varieties. The Canna is not only a beautiful plant, but also a useful one; from the tubers of some of the species, especially from *Canna edulis*, known also under the name of *Tous les mois*, an excellent kind of arrowroot is prepared; the tubers of others are used as a vegetable and the leaves are used for packing. The Cannas just now most popular are the dwarf varieties produced from species less showy, by the famous French gardener Crozy; it may indeed be said that there exists at present a Crozy craze among the florists and Canna fanciers. The Conservatories possess a fine suite of those Crozy Cannas and whoever examines their gorgeous flowers will admit that they richly deserve their popularity. The suite includes the following varieties:

Madame Crozy, flowers, scarlet, bordered with gold.

Alphonse Bouvier, large crimson flowers.

Paul Bruant, brilliant orange flower merging into scarlet.

J. D. Caboz, flowers orange, with a pinkish tinge.

Eldorado, flowers golden yellow.

Egandale, foliage dark maroon-green; flowers, bright cherry color.

Charles Henderson, flowers, scarlet merging into crimson.

Sarah Hill, flowers, red.

Paul Marquant, a large and beautiful flower the color of which is not easy to define; it might be called a bright salmon with a tinge of carmine.

Beauty of Poitevin, a glowing dark red flower.

Capitaine P. de Suzzoni, yellow flowers mottled with scarlet.

Florence Vaughan, lemon yellow spotted with bright red.

THE MARANTAS OR ARROWROOT.

These foliage plants, *par excellence*, find few equals in beauty and peculiarity of design of the leaves and in richness and variety of color. They are named after Balthazar Maranti, an Italian botanist, and are found in the tropics of both hemispheres. From the fleshy root-stalks of some of the species the starchy matter known as Arrow-root is obtained. Some of the plants called Maranta by the horticulturists are really Calatheas, a closely allied species. (*Kalathos*, a basket, in reference to the leaves being worked into baskets in South America.) The name Maranta is retained in the following list because the plants are thus labeled and popularly better known under that name. The species properly belonging to Calathea are marked (c).

The following species can be found in the conservatories:

Maran'ta bi'color, two-colored Maranta. Has round leaves of pale green color with irregular blotches of dark olive; underneath, rose purple. From Brazil. [s.]

M. Chimboracen'sis (Chimborazo). Ground color, light green with zigzag markings of olive green, bordered with white. An elegant species from Ecuador. [s.]

M. distich'ium (two ranked). [A.]

M. (c.) exim'ia (choice). [s.]

M. (c.) fasciat'a (banded). [s.]

M. hirsut'a (hairy). [s.]

M. (c.) Kerchowian'a (Kerchow's). Leaves green with two rows of brown blotches. From Brazil. [s.]

M. (c.) Leit'zei (Leitze's). Leaves deep green, with short yellow-green bands. From Brazil. [s.]

M. (c.) Lenden'ii (Linden's). Leaves banded with pale and dark green. From Peru. [s.]

M. (c.) **Makoyam'a** (Makoy's). One of the largest and handsomest species and which will well bear close examination; what a cunning artifice of nature to design upon the large blade of the leaf a branch composed of alternating large and small leaves, green above red on the underside, and this branch is painted on a background that resembles a woven fabric. [A. & S.]

M. (c.) **Massangean'a** (Massange's). From Brazil. [A. & S.]

M. (c.) **Mic'ans** (glittering). From Brazil. [S.]

M. **Por'teana** (Porte's). Bright green on the upper side, striped with transverse lines of white, purple beneath. From Bahia. [A. & S.]

M. (c.) **Prin'ceps** (Prince). From Peru. Leaves metallic green with two yellow bands, purple beneath. [A.]

M. (c.) **pulchel'la** (pretty). From Brazil. Leaves bright green with two series of deep green blotches, alternately large and small. [S.]

M. **regal'e** (royal). Dark green leaves with double parallel stripes of carmine. [S.]

M. (c.) **rosea-pict'a** (rose-colored). From upper Amazon. Leaves rose banded, purple beneath. [A. & S.]

M. **sanguin'ea** (bloody). [A.]

M. (c.) **Seeman'ni** (Seemann's). From Nicaraugua. Leaves satiny emerald-green, midrib whitish. [S.]

M. **Smaragdin'a** (emerald green). From Ecuador. Leaves emerald green with a dark green central stripe. [A.]

M. (c.) **splen'dida** splendid). From Brazil. Leaves green, banded purple beneath. [S.]

M. **Spitzerian'a** (Spitzer's). [S.]

M. c. **Van'den Heckei** (Van den Heck's). From Brazil. Leaves deep green marked with gray. [A. & S.]

M. (c.) **Veit'chii** (Veitch's). From Western Tropical America. Leaves green blotched. [A. & S.]

M. (c.) **Virginal'is** (Virginal). From the Amazon. [A. & S.]

M. (c.) **Warzewic'zii** (Warzewicz's). From Tropical America. [S.]

M. (c.) **Zebrin'a** (zebra-like.) From Brazil. Leaves alternately dark and light green striped. [A. & S.]

THE DRACÆNAS.

These handsome foliage plants are too well known and too popular to need an introduction to the reader; they belong to the Lily family and are distributed over nearly the whole tropical world, forming many species to which cultivation has added many varieties. Only one of the species, Dracæna Draco, is recognized by the botanists of to-day as being the Dracæna proper, all the others have been placed into different genera; but the horticulturist and the public are still using the older designation which is therefore retained here.

Dracæn'a Drac'o. The Dragon tree. (*Drakaina*, a female dragon; the plant owes the name to the fact that, if it is wounded, a milky juice flows from the wound which on drying becomes a hard gum, having similar properties as the resinous substance called Dragon's Blood.) This is the tallest species of the Lily tribe with straight, column-like trunks which are surrounded from bottom to top with large dark leaves. The branching spikes of the rather small, liliaceous flowers can often be seen set obliquely at or near the top of the plant. The Dragon tree is famous from the immense specimen which grew on the island of Teneriffa, and was described in 1402 by the French adventurer Bethencourt, and 400 years later by Humboldt. It was blown down by a hurricane in 1867; its age has been calculated as 5,000 years. [A. & S.]

Dracæna amab'ile (amiable), a very pretty form with the upper leaves variegated with pink and white. [A. & S.]

D. australis (Southern). [A.]

D. Baptist'ii (Baptist's). [A.]

D. bel'la (beautiful). Leaves almost black. [S.]

D. Brazilien'sis (Brazilian). [A. & S.]

D. Bruantii (Bruant's). [S.]

D. Dennison'ii (Dennison's). Has large, dark red leaves, the upper ones being brighter. [S.]

D. ensifol'ia (sword-shaped leaves.). [A. & S.]

D. frag'rans (fragrant). From Africa. [A. & S.]

D. Goldiean'a (Goldie's). Leaves greyish with dark green transverse bands. South Africa. A very magnificent species. [A. & S.]

D. grac'ilis (graceful). [A.]

D. Guilfoy'lei (Guilfoyle's). [A.]

D. hy'brida (hybrid). Bright leaves, colored pink and green.
[A. & s.]

D. indivis'a (undivided). [s.]

D. Linden'ii (Linden's). Yellow leaves with green along the midrib. [A. & s.]

D. porphyrophylla, leaves marked like porphyry. [A.]

D. Seiboldii (Seibold's). [A.]

D. Shephard'ii (Shephard's). A large plant with purple edged leaves. [A. & s.]

D. spectab'ilis (showy). [s.]

D. stricta (straight). The leaves are dark red. [s.]

D. terminal'is (terminal). A fine plant with crimson leaves; from the East Indies. [A. & s.]

D. umbracolif'era (umbrella-bearing). The leaves are dark green and very closely set; they spread out horizontally with their edges turned down, giving the plant an umbrella-like appearance. From the island of Mauritius. [s.]

D. Young'ii (Young's). A fine species with dark red leaves.

Some other plants of the Lily family may well be grouped here; such as

Aspar'agus plumos'us nan'a (*Sparasso*, to tear; on account of the strong prickles of some species which are fatal to clothes; *plumosus*, feathery; *nana*, small). From South Africa. [A. & s.]

Asparagus tenuis'simus (very narrow leaved). These two are very graceful relatives of the asparagus which we buy in the market (Asparagus officinalis). Our florists use them in great quantities to interweave them in floral pieces and they are generally taken for ferns by the people. [A. & s.]

Phorm'ium ten'ax, New Zealand Flax. (*Phormos*, a basket, the fibers being used for making baskets; *tenax*, tough.) This interesting plant from New Zealand has sword-shaped leaves which grow in opposite rows and clasp each other at the base. The large flower spikes rise from the center of the leaves and attain considerable height. The leaves contain a large quantity of strong fiber which is put to many uses by the natives; but its separation from the plant is somewhat difficult, for this reason the New Zealand flax has not attained yet any prominence as an article of export.

GRASSES.

Some Grasses and Sedges are well adapted to be used as foliage plants, indeed there are some of our own wild grasses which deserve attention for their natural beauty; those we find in the conservatories are the following:

Dact'ylis glomerat'a, Cock's-foot grass. (*Daktylos*, a finger) A grass which grows wild in this country and in Europe and which is considered pretty enough to be cultivated in greenhouses.

Gymnos'tichum Pier'cei (*Gymnos*, naked; *stichos*, rank). This is a cultivated species of the Bottle-Brush grass.

Pan'icum sulcat'um (probably from panis, bread).

Panicum variegat'um (variegated). This is an especially pretty grass with leaves striped green and white and tinted pink; an elegant plant to grow in baskets. (The Sedges will be mentioned among the aquatic plants).

SPIDERWORTS.

The Spiderworts are raised principally for their lustrous and prettily variegated foliage; they need little care and often grow only too rapidly, spreading over rockeries or running over neighboring flower pots, striking root wherever there is the slightest chance, from which habit they have received the name "Wandering Jew." The botanical name, Tradescantia, was given in remembrance of J. Tradescant, gardener to Charles I. The following species are found in the Conservatories, most of them among the rocks in Schenley Park Palm house.

Tradescant'ia dis'color (Two colored, the leaves being green above and purple beneath). This plant would not be taken at the first glance, to be a Tradescantia, its rigid, spirally arranged leaves resembling those of some Bromelias, but the flowers which, though not very conspicuous, peep curiously out from flattened bracts in the axils of the leaves, betray their relationship. It grows wild in South America.

T. fuscat'a (*fuscous*, refers to the dark hairy covering of the stem). Also from South America.

T. repens (creeping).

T. specios'a (showy). From Mexico.

T. velut'ina (velvety) and

T. zebrin'a, striped with light green and reddish purple. From South America.

T. Virgin'ica. Virginia Spiderwort, is the species so much grown in gardens on account of its pretty blue flowers; it is found growing wild below Pittsburg along the Ohio river and is very common in the South.

Dichorisan'dra musa'ica (*Dis*, twice; *chorizo*, to part; *aner*, anther; *musaica*, mosaic). This is another plant of the Spiderwort family; its leaves are prettily marked and veined with yellow or white upon the dark green ground, while the under side is deep reddish purple. The flowers are of a beautiful azure blue.

PINEAPPLE FAMILY.

Several interesting plants of the Pineapple family are cultivated on account of their brilliantly colored leaves, some also on account of their handsome flowers. The most interesting to many visitors is

Ananas'sa sativ'a, Pineapple. (*Nanas*, the name of the Pineapple in its home in Peru; *sativa*, cultivated.) The delicious flavor of the fruit of this plant has long ago been written about by travelers to South America. The first knowledge of it came to Europe in 1558 through a monk who had visited Peru. A Huguenot priest, Jean de Lery, described it three hundred years ago as being of such excellence that the gods might luxuriate upon it and that it should only be gathered by the hands of Venus. The Pineapple is a biennial, ripening and finishing its growth in the second year. The fruit is produced on a short stem rising from the middle of a large rosette of leaves while the top of this stem is also crowned with rigid, spiny leaves. The fruit is the product of many flowers whose ovaries grow so closely together that they appear as one single cone-like mass. The Pineapple is now extensively cultivated not only in South and Central America and the West Indies but also in India and the East Indian Islands. Both conservatories.

Billberg'ia zebrin'a (Named after the Swedish botanist, Billberg; *zebrina*, zebra-streaked). A plant with rigid leaves and elegant yellow or reddish, fragrant flowers. In their home in tropical America they grow upon trees. They are favorites with the people in South America and can be often seen hanging in windows or on balconies. [A.]

Pitcairn'ia corallin'a. Coral plant. (Named after Dr. Pitcairn; *corallina*, coral-red). This handsome plant, with long, narrow spiny leaves, produces a long spike of intensely scarlet flowers which has been compared with a branch of red coral ; it is a native of Columbia. [A. & S.]

Tilland'sia utriculat'a. Air plant. (Named after Dr. Tillands ; *utriculata*, bladdery.) This is one of the most interesting of the air plants; its leaves have a bottle-like cavity at the base which is capable of holding a considerable quantity of water, a most welcome provision for the thirsty wanderer ; it inhabits trees in the f rests of Jamaica. Another plant of this genus, Tillandsia usneoides, is known to most readers under the name of "Florida Moss ;" hanging from the trees in rich, silvery-gray, lace-like festoons, it looks indeed more like a moss than a relative of the bulky Pineapple ; but its flower betrays its botanical relationship. This Spanish moss is collected in great quantities and it furnishes excellent packing material ; it is also used as a substitute for horsehair.

Tillandsia cyane'a (blue). This species and many others, for there is a great number of species of this genus, have thickish, strap-shaped leaves, growing in the form of a rosette from the center of which the flower-stalk arises, bearing a number of blue flowers. From Guatamala. [A.]

Tillandsia zebrina also called *splendens*, is a handsome plant with stout leaves, cross-striped with alternate bands of light green and reddish brown. From French Guiana [A. & S.]

Tillandsia tessellat'a has, as the name implies, tessellated or checkered leaves of green and yellow color. A synonym is Vrisia tassellata. From Brazil. [A.]

Vries'ia musaic'a (named after Dr. De Vries, a Dutch botanist ; *musaica*, mosaic). The Vriesias are closely related to the Tillandsias ; they have flat leaves and bear the flowers in spikes, protected by large, handsomely colored bracts. [S.]

Æchme'a Mariæ Regin'æ (*aichme*, a point, on account of the rigid points on the flower envelopes). A very handsome plant with rather long and stout leaves, arranged in a loose spiral rosette and the tips gracefully turned downward ; the large and showy flower-spike is closely covered with blue-tipped flowers. This species, the finest of the Æchmeas, is dedicated to Queen Mary. From Costa Rica. [A.]

Nidular'ium spectab'ilis (*nidus*, a nest; *spectabilis*, showy). The name refers to the nest-like rosette of the thickish leaves; in the center of the rosette the leaves are tipped with bright red, giving the plant the appearance as if there were a flower in the middle. At blooming time a flower-stalk rises from the centre, bearing red flowers. These plants are also called Karatus. From Brazil. [A.]

THE GINGER PLANTS.

Zing'iber officinal'e. Ginger. (*Zingiberis*, the Greek name, derived from the Sanskrit, in which language it means horn-shaped; *officinale*, commercial). Every visitor is interested in this plant which furnishes such a peculiar spice of highly pronounced and characteristic taste and flavor. The plant is not very showy; the leaves are arranged in two ranks on the stem, which they clasp with their sheathing bases; the flowers are borne on cone-shaped spikes, thrown up from the rootstalk and protected by bracts. A peculiarity of the stamens is that the filaments reach beyond the anthers in the form of a beak. The most important part of the plant, however, is not visible; it is the creeping underground stem or rhizome which yields the ginger of commerce. The plant is largely cultivated in the East and West Indies as well as in Africa and China. The "Best Preserved" ginger comes from the West Indies, that from China ranks next. The rhizomes, commercially called races, are prepared when the plant is about one year old; they are cleaned and dried; this is the ginger root we buy in the stores. With the various preserves, decoctions and beverages made from it the reader is probably well acquainted.

Cost'us Malortiean'us (*Costus*, is the ancient name; *Malortieanus*, Malortie's). This beautiful plant belongs to the same family as the Marantas and Bananas; it has rich, soft green leaves with fine parallel veins and when blooming bears a golden yellow flower.

[S.]

Amom'um cardamom'um. Cardamom. (From *a*, not; and *momus*, impurity; because it was considered to counteract poisons and prevent decay.) These aromatic herbs were used in embalming, whence the word mummy. The cardamom seeds used for flavoring wines and spirits are derived from this and other species. Native of India.

Hedych'ium coronar'ium (*Hedys*, sweet; *chion*, snow; in reference to the sweet scented snow-white flowers; *coronarium*, a crown or garland). From the East Indies, a beautiful and fragrant flower.

Curcul'igo recurvat'a (*Curculigo*, a weevil; the seeds have a beak like that beetle; *recurvata*, turned-back leaves). This foliage plant with large grass-green, plaited leaves has been mentioned before; it comes from Bengal and thrives only too well in the greenhouse. It belongs to the Hypoxis family. Allegheny Conservatory has also a variegated specimen of this genus.

ACANTHUS PLANTS.

Many plants of the Acanthus family are distinguished by their square stems and beautiful foliage; some are most elegant in shape while others are exquisitely colored. One of the species, Acanthus mollis, with most ornamental foliage, is said to have suggested the leaf ornament on the Corinthian column. Acanthus means spiny, some of the plants of this family being thistle-like in aspect.

Sanchez'ia nob'ilis variegat'a (named after Joseph Sanchez, Professor of Botany in Cadiz). This beautiful plant can be seen in the picture representing a group of foliage plants; the yellow veins and midrib in the bright green leaf make it very attractive; it bears yellow flowers and is a native of Ecuador.

Sanchezia macrophyl'la, (long-leaved) is another fine specimen recently brought into cultivation.

Strobilan'thus Dyerian'a (*Strobulus*, a cone; *anthos*, a flower; on account of its cone-shaped flower-spike. Named after Dr. Dyer of the Kew Gardens in London). Do not fail to examine the peculiar tints of the leaves of this plant; dark to light lavender with green veins, a combination, probably found in no other leaf.
[s.]

Fitton'ia Pearc'ei (named in honor of S. M. F. Fitton, author of "Conversations in Botany"). This plant with broad, bright green leaves and carmine veins is one of the prettiest of the small foliage plants. Not less attractive are

Fittonia argyneur'a (silver-nerved) with silvery-white veins and

Fittonia macrophyl'la (long-leaved.)

Ruel'lia Devosian'a (named in honor of Jean Ruelle, botanist and physician to Francis I.). This plant has white flowers and most exquisitely marked leaves, the ground color being a rich velvety green, with whitish veins and purple beneath.

Meyen'ia erect'a (named after M. Meyen) more commonly called

Thunberg'ia erect'a (named after C. P. Thunberg, Professor in Upsala, who had visited Batavia and Japan). This genus contains many attractive plants a number of them climbing or twining and well adapted for floral baskets. The above species produces its beautiful flower nearly throughout the whole year. The corolla is deep blue with orange throat and yellow tubes.

Eran'themum pulchel'lum (From *erao*, to love; *anthos*, a flower; *pulchellum*, pretty). Indeed one of the prettiest of this genus, with blue flowers.

Peris'trophe angustifol'ia (*Peristrophe*, turning round; because the corolla is twisted so as to be upside down; *angustifolia*, narrow-leaved).

Thyrsacan'thus rut'ilans *Thyrse*, a dense, pyramidal flower cluster; *acanthus*, see above; *rutilans*, reddish). A handsome plant with bristle-pointed leaves and brilliant crimson flowers. Also known under the name Thyrsacanthus Schomburgkianus.

The last five species should be placed more properly in the chapter on flowers, but are mentioned here in order not to separate them from the other plants of the Acanthus family.

Grevil'lea robust'a (named after C. F. Greville, a patron of Botany.) A very popular and easily grown foliage plant with finely dissected and compound leaves, resembling some fern fronds. The Grevillias are natives of Australia and New Caledonia. They belong to the family Proteacea.

Peperom'ia argyr'eya (*Piper*, pepper; *omoois*, similar, being closely related to the pepper plant; *argyneura*, silver veined). The Peperomias are well known and easily cultivated plants, admired for their shield-shaped, fleshy and prettily marked leaves. The above species has leaves which are bright green along the radiating veins while the spaces between are of a silvery gray.

Peperomia microphyl'la (small-leaved). This is a Mexican species with small, usually whorled leaves. The Peperomias belong to the Pepper family or Piperaceae.

Sphaerog'yne latifolia, Turtle-back Plant. (*Sphaira*, a globe; *ryne*, female, on account of the globular ovary; *latifolia*, wide-leaved.) This handsome foliage plant with peculiar and conspicuous venation and its rich color, dark green above and crimson purple beneath, is not easily overlooked by the visitor; its home is in Guiana, South America, where it is known under the name of Tococa. It belongs to the family of Melastomaceae. To the same family belongs

Cyanophyl'lum magnif'icum (*Kyanos*, blue; *phyllon*, leaf). This is one of the grandest ornamental-leaved plants cultivated in the hot houses; its large, five-nerved leaves are of a rich shaded green above and purple beneath. It comes, like the above, from tropical America.

Nepeta glech'oma variegata, Ground Ivy. (*Nepeta*, an old Latin name, used by Pliny, probably derived from the Italian town of Nepi.) What child does not know that everywhere present little vine, the Ground Ivy, a common weed in many countries and still it has been found worthy of a place among the rockeries of Schenley Park Palm house, simply because some gardener succeeded in causing it to produce a mottled leaf, green and white; but it fills its assigned place very gracefully and for making pretty festoons over little projections of the rocks it is just the thing. It belongs to the Mint family.

THE CROTONS.

These are plants of the Spurge or Euphorbia family and some of them, as Croton tiglium, are known as the source of a valuable, but unpleasant drug, the Croton oil, which, I hope, the kind reader will never be called upon to take. Croton Eluteria furnishes another drug, the Cascarilla bark. The word Croton means a tick and refers to the appearance of the seed. The Crotons proper furnish few species worthy of cultivation; the plants so numerously found in most of our green houses belong to the closely allied genus Codiæum. Still as Croton is the name used by the florists as well as by nearly all other people except botanists, it is retained here.

The influence of cultivation and artificial stimulation of plants to induce them to vary from their natural form is in few cases more strikingly illustrated as in the Crotons. The natural species of this genus which are found widely distributed in the tropics, are only attractive on account of their thrifty, shiny, leathery foliage with a tendency to vary from their dark green to rich hues of brown, red and yellow. Cultivation by planting seeds and cuttings, selecting such with prominent new features of color and form, has led to the production of many varieties, distinguished from each other by difference in color, markings and shades, difference in the shape of the leaf from the simple outline to variously lobed margins, from smooth and flat blades to such with crimped edges and completely spiral twisted forms. Thus the Crotons have become favorite foliage plants and they are interesting as they show how man can become creative if he studies nature closely and follows her hints.

The following is a list of all the Crotons cultivated in the Conservatories at the time of this writing; they are mostly varieties derived from a few species and the majority may be referred to Codiæum pictum. The name Codiæum is derived from Codebo, the Malayan name of one of the species.

CROTON (CODIÆUM).

Andraean'um.	[A. & S.]
Aucubæfol'ium (Aucuba-leaved).	[A. & S.]
Aneitinnen'se.	[A.]
Aurea-maculat'a (golden spotted).	[S.]
Bergmann'ii.	[S.]
Bis'marck.	[A.]
Caudat'us-tor'tilis (twisted-tail).	[S.]
Chal'lenger.	[S.]
Cornut'um (horned).	[A.]
Compte de Germiny.	[S.]
Cronstadt'ii.	[S.]
Earl of Derby.	[S.]
Eburn'eum (ivory-white).	[A.]
El'egans.	[A.]
Evansian'um.	[S.]
Fasciat'um (bundled.	[S.]

Gold'iei.	[A.]
Hanburyan'um.	[S.]
Hillean'um.	[A.]
Hookerian'um.	[A. & S.]
Inimitab'ilis.	[S.]
Interrupt'um.	[A. & S.]
Irregular'e.	[S.]
James'ii.	[S.]
Johannes.	[A.]
Lady Zetland.	[S.]
La Dame Blanche.	[A.]
Lancifol'ium.	[S.]
Lord Cairns.	[A. & S.]
Macarthur'ii.	[A.]
Magnif'icum.	[A. & S.]
Makoyam'a.	[S.]
Majes'ticum.	[A. & S.]
Max'imum.	[A.]
Morean'a.	[A.]
Mort'ii.	[A. & S.]
Multi-color (many-color).	[A. & S.]
Nes'tor.	[A. & S.]
Nobil'is.	[S.]
Ovalifol'ium.	[S.]
Pict'um (painted).	[a]
Picturat'um (painted).	[A.]
Prince of Wales.	[S.]
Proemors'um.	[A.]
Queen Victoria.	[A. & S.]
Regale.	[S.]
Rex.	[A.]
Rosea-picta.	[S.]
Spiral'is.	[A. & S.]
Splendens.	[S.]
Sunbeam.	[S.]

Trilob'um (three-lobed).	[A.]
Undulat'um (wavy).	[A. & S.]
Variabilis.	[S.]
Variegat'um.	[S.]
Veit'chii.	[A. & S.]
Victory.	[A.]
Warren'ii	[S.]
Weisman'nii.	[A. & S.]
Young'ii	[A. & S.]

As will be seen, most of these forms are named after persons and the meaning of the other species names is so evident that no definition is necessary. Some of the characteristic leaves are shown in the adjoining picture.

List of Croton Leaves Represented in the Illustrations on Page 75.

1. Croton Cronstadtii.
2. " Hookerianum.
3. " Rosea picta.
4. " Challenger.
5. " Earl of Derby.
6. Croton Queen Victoria.
7. " Compte de Germiny.
8. " Mortii.
9. " Splendida.
10. " Aurea maculata.

THE GREATEST CHAIR ON EARTH.

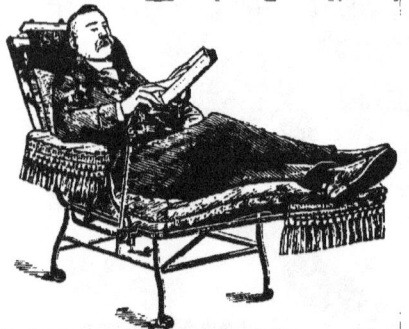

. STEVENS' IMPROVED .
Adjustable . .
Folding Chair

HAS NO RIVAL !

Everybody's Favorite.
50 Changes of Position.
5 Articles in One.

Parlor CHAIR or a full length Bed. An adjustable Lounge or Couch. A Child's Crib or an Invalid's Chair. The most useful Chair ever devised. Its Merits: Simplicity, Durability, Elegance and Comfort. Prices at discount to suit the times. Also, **Wheel Chairs**. All kinds Porch, Lawn and Hammock Chairs.

Address **Stevens Chair Co.** No. 3 Sixth Street. PITTSBURGH, PA.

Chickering, Hardman
    Kimball   } **PIANOS**.
Krakauer, Vose

Kimball Pipe
Peloubet Pipe
United States   } **ORGANS**.
Chicago-Cottage
Edna

PIANOS . .
To Rent, Repaired, Polished, Tuned, Carted, Stored at most Reasonable Rates.

CASH OR INSTALLMENTS.

Many Second-Hand Pianos, Square and Upright, at Great Bargains. Old Instruments taken in exchange for New Ones.
Circulars Free to any Address.

MELLOR & HOENE,

ESTABLISHED 1831.

Warerooms. 77 FIFTH AVENUE, PITTSBURG, PA.

Largest and Oldest Music House in this part of the U. S.

SPROUTING COCOANUT PALMS.
The nuts are from Mr. Charles J. Clark's Cocoanut Plantation in Florida.

ORCHIDS.

VANDA TRICOLOR VAR. PATTERSONI,
Allegheny Conservatory.
(See page 90)

ORCHIDS.

Of all strange and odd plants the oddest are the Orchids, and with all their oddity they are beautiful. They are fashionable now as a flower for the personal adornment of our fair ladies and they are becoming more so, and the young admirer who can afford to present his admired one with a spray of orchids, is pretty sure of a sweet smile of gratitude. With the flower fanciers the Orchids take the place now which the tulips and the roses took in former times; owners of Orchidaries, if I may be permitted to coin such a word, are vying with each other to get the choicest and rarest collection, and high prices are paid for new species and striking varieties. The rare tulips, for the possession of which fortunes were offered and intrigues were spun, were the result of skilful breeding by the cunning Dutch florist; the Orchids, however, which bring the highest prices, are new species discovered by the Orchid hunter.

A peculiar, fascinating and not seldom dangerous occupation it must be, that of and Orchid hunter. He must penetrate the darkest recesses of the Brazilian Selvas, to scan with his field glass the highest branches of the trees in order to discover the beautiful epiphytes (air plants); for many of these plants, not being able to get sufficient sunlight and pure air on the darkly shaded ground, have made themselves at home in the lofty crowns of the trees, where they seem to sit, turning their butterfly-like flowers toward the blue sky, while the long, fibrous roots hang down like long fringes, absorbing the moisture which rises from below. During the rainy season they are rich in foliage and produce their wondrous flowers, but during the dry season they often shrivel into a heap of apparently dead sticks and branches; life, however, is not extinct; many species have thickened joints of stems, so called pseudo bulbs (false bulbs) at the base, which are able to contain moisture and nourishment for a long time and which give the plant a vigorous start as soon as the drouth is over. Even after such a prize is discovered, it is often with the greatest difficulty and at the risk of life and limb that the hunter can obtain it; he must be an expert tree climber to take the Dendrobioms, Epidendrums, Lælias, Cattleyas from their lofty seats; he must be a venturesome mountaineer to wend and work his way over crag and cliff in the wildest part of the Andes mountains where the Lycastes, Masdevallias, Oncidiums and Anguloas have their abode; he must have nerve to penetrate the dark and darkest Africa, where rare and new forms may be his reward. The islands of the Indian ocean, India and the Malay peninsula, though much explored, may still harbor some strange and unknown species.

The plants collected are carefully packed in moss and sent to England, where the packages are eagerly examined by the consignees. Alas, too many of the precious plants do not survive the voyage; especially of those who had to be carried for days and weeks on mule-backs before reaching a railroad or the seaboard. But if only a spark of life is apparent, the greatest care is taken to save the specimen.

While we enjoy the sight of the beautiful Orchids, we cannot help thinking it a great pity that so many of the lovely plants should be ruthlessly torn from their native soil and sometimes none left to reproduce the species in its home. It is a fact that some of the species cultivated in conservatories are no longer found in a wild state. Even some of our own native Orchids, such as Cypripedium spectabile, the Showy-Lady-Slipper, formerly plentifully found in marshy places in Pennsylvania, is getting scarce because florists pay a good price for the roots.

But why should the plants of this family be so highly prized and so much sought after? Partly, no doubt, on account of their rarity and their peculiar habits of growth; but not less on account of the strange forms of their flowers and their beautiful colors.

The ground plan of the flower of the Orchid is not different from that of other flowers of the endogenous class. There is a perianth or floral envelope of six divisions, three of them may be called the calyx and are generally alike in shape and color; the other, inner, three represent the calyx and of these one is always different from the other two; often very much so in color as well as in shape; it is called the lip or labellum and has sometimes on its posterior end a tube or spur called the nectarium, because it generally contains honey. In the accompanying pictures of Orchids as well as in the specimens in the Conservatories, when in flower, the reader will have no difficulty to distinguish these parts. The lip may be found on the upper and back part of the flower or it forms the lower and front part; in the latter case it will always be found that the flower-stalk or pedicil has a twist, giving the flower a half turn and thus bringing the lip to the front. Instead of having six stamens as the lily and many other of that class, the Orchid has only one, in the Cypripediums two—and it is attached to the style of the pistil, forming with it the "column"; the stigma is on this column below the stamen and the arrangement is such that it is impossible for most Orchids to fertilize themselves, that is to spread the pollen upon the stigma. But if an insect, a wasp for instance, visits the plant to get honey, the two pollen bags of the stamen, being provided with a sticky disk, stick to its head; the wasp flies off with its unwelcome ornament, which looks like two plumes, and on entering the next flower, those bags, having meanwhile changed their positions, run right against the stigma, which with its sticky surface, tears the mass and becomes provided with the fertilizing pollen.

If any reader should like to learn more about this wonderful arrangement, he should read Darwin's book "On the Fertilization of Orchids by Insects."

We are not without Orchids in this country; our Moccason plants or Lady Slippers, Cypripedium acaulis, and Cypripedium pubescens, both of which may be found in this county, are scarcely inferior to those generally found in green houses and how Cypripedium spectabile is sought after has already been mentioned; besides, we may find Orchis spectabilis in our woods and sweet-scented Ladies' Tresses (Spiranthes) in moist meadows as well as other less conspicuous members of the family.

As useful plants the Orchids do not rank high. A drug called salep is prepared from the starchy root of several species; it is used as a nervine restorative and fattener; but much more important is the climbing Orchid Vanilla aromatica and planifolia, the plants which yield the pod, called Vanilla bean, furnishing the most delicious flavoring substance known, as young lovers of vanilla ice cream will gladly testify.

The species of Orchids are almost innumerable, between 4,000 and 5,000 are named and known; the botanists divide them into a number of tribes of which only those will be mentioned here which have representatives in the in the Conservatories.

Mr. Hamilton of Allegheny Conservatory prides himself on his large and fine collection, and well he may, for it contains many and most beautiful forms. While the majority bloom in winter and early spring, the season which used to be summer for them when they were at their home in the Southern hemisphere, visitors can find some fine specimens in flower whenever they take a walk through the narrow, but lovely passage of the Orchid house and some habitués visit this department every week in order to enjoy every flower as they unfold one after another. Schenley Park Conservatory has a good collection and more will be added as means and opportunity will permit.

To describe every species would take a book alone ; only a list can be given here with casual remarks.

TRIBE EPIDENDRAE.

TREE ORCHIDS.

The name of this tribe is derived from two Greek words, *epi*, upon ; and *dendron*, a tree ; because most of these orchids grow on trees; they are, however, not parasites, taking their nourishment from the substance of the tree, but epiphytes, using the branches of trees simply as support and getting their nourishment from the air.

Many of the finest Orchids belong to this tribe which is divided into several genera.

Epiden'drum. The genus of this name contains over four hundred species, most of them small and medium size. The lip of the flower has a spreading limb (upper part) and its claw adheres to the column, the base of which has a long, deep hollow. They are found in the forests of South America, Central America, Mexico and the West Indies. The following are some of the finest species :

Epiden'drum aloifolium (aloe-leaved).	[A.]
E. aur'eum (golden).	[A.]
E. caloch'ilum (beautifully lipped)	[A.]
E. ciliar'e (fringed).	[A.]
E. cinnabar'ium (cinnabar-red).	[A.]
E. cochleat'um (spiral).	[A. & S.]
E. falcat'um (sickle-shaped).	[A.]
E. frag'rans (fragrant).	[A.]
E. Vittelin'um (colored like the yelk of an egg).	[A.]

Cattleya (named after Mr. Cattley a distinguished patron of botany). To this genus belong many of the largest and handsomest Orchids which have justly become popular and are cultivated in all first class nurseries. The Cattleyas generally have the lowest joint of their stems thickened into a pseudo-bulb; sometimes this joint is club-shaped. Some have a single leaf on the top of each stem, others have two or three. The flowers are scarcely surpassed by any other Orchid in size and brilliancy of color. They rise from the top of a pseudo-bulb and are enclosed in a sheath. The stamen has four pollen masses.

Cattley'a Aclan'diae.	[A.]
C. amethys'tina (amethyst-colored).	[A.]
C. amethystoglos'sa (amethyst-tongue).	[A.]
C. Alexandra.	[A.]
C. aur'ea (golden).	[A.]
C. Bouringian'a.	[S.]
C. Chocoen'sis.	[A.]
C. citrin'a (citron-flowered).	[A.]
C. Dowian'a.	[A.]
C. Eldorad'o.	[A.]
C. Gaskellian'a.	[A.]
C. gigas (gigantic).	[A.]
C. granulos'a (granulated-lipped).	[A.]
C. guttat'a (spotted).	[A.]
C. guttata Leopoldii.	[A.]
C. intermed'ia (intermediate).	[A.]
C. labiat'a (lipped).	[A. & S.]
C. Leean'a.	[A.]
C. max'ima (largest).	[A.]
C. Mendel'ii	[A.]
C. Mos'siae.	[A. & S.]
C. Mossiae Victoriae.	[A. & S.]
C. Percivalian'a	[A.]
C. Schillerian'a.	[S.]
C. Skin'neri.	[A.]
C. Trian'ae.	[A.]

CATTLEYA SKINNERI.
(Orchid)
Allegheny Conservatory.

Cattleya, Trianae formosa. [A.]
C. Walkerian'a. [A.]
C. War'neri. [A.]
C. Warscewic'zii delicata. [A.]

Lælia, the vestal virgin, has been given a place among the Orchids; no doubt the loveliness and delicacy of the flowers of this group has suggested that name. The Lælias are closely allied to the Cattleyas, from which they are distinguished by having eight pollen masses to their stamens. They are found from Mexico to Brazil.

Læl'ia anceps (two-edged . [A. & S.]
L. autumnal'is (autumnal. [S.]
L. acuminat'a (pointed-lipped). [A.]
L. Dayan'a. [A]
L. peduncular'is. [A.]
L. Perrin'ii. [S.]
L. purpurat'a. [A. & S.]

Dendrobium; this name means also living upon a tree; the plants of this genus are distinguished by their tall, slender, jointed stems, often resembling a reed; some have pseudo bulbs. Some species produce large, delicately tinted flowers and some have a delicious odor. The lip is more or less contracted into a claw, lying upon or attached to the front of the column. They are natives of India, Japan and Australia.

Dendrob'ium bi-gib'bum, double spurred. [S.]
D. Bullerian'um. [A]
D. calceolar'ia, slipper-like. [S.]
D. calceol'us. [A.]
D. Cambridgian'um, Duke of Cambridge. [A.]
D. chrysan'thum, golden flowered. [A.]
D. chrysoton'um, golden-arched. [A.]
D. cœrules'cens, sky-blue. [A.]
D. crassinod'e, thick-knotted. [A.]
D. Dalhousian'um. [A.]
D. Dear'ei. [A.]
D. densiflorum, dense-flowered. [A.]

D. Devonian'um. [A.]
D. Far'meri. [A.]
D. Falconer'i. [A.]
D. fimbriat'um, fringed. [A.]
D. formosum-gigan'teum, beautiful, large. [A.]
D. infundib'ulum, funnel-shaped. [A.]
D. Jamesian'um. [S.]
D. moschat'um. [S.]
D. nob'ile. [S.]
D. Pharish'ii. [S.]
D. Phalænop'sis. [A.]
D. Pierar'dii. [A.]
D. Schroed'eri. [S.]
D. Thyrsiflor'um. [A. & S.]
D. Wardian'um. [A.]

Barker'ia el'egans (named after G. Barker an ardent cultivator of Orchids). From Mexico.

B. Skin'neri.

Blet'ia grandiflor'a (named after Don Louis Blet a Spanish botanist).

B. Shepher'dii.

B. Tunkervil'lia.

These are terrestial Orchids from the West Indies.

Broughton'ia sanguin'ea (named after the English botanist, Broughton; *sanguinea*, blood-red). From the West Indies.

Brassavol'a glauc'a (named after Brassavole, an Italian botanist).

B. Sanderian'a.

B. acaul'is. These are from Central America.

Coelog'yne flac'cida (from *koilos*, hollow; *gyne*, female, in reference to the pistil; *flaccida*, drooping). [A. & S.]

C. cristata al'ba (white-crested). [A.]

C. specios'a, showy. [S.]

They are natives of India, Southern China and the Malayan Archipelago.

Chys'is bractes'cens (from *chysis*, melting, from the fused appearance of the pollen masses; *bractescens*, having bracts). From Guatamala.

Masdeval'lia maculat'a (named after J. Masdeval, a Spanish botanist; *maculata*, spotted). The Masdevallias are small epiphytes, remarkable for the singularity of their flowers; they are found at considerable heights in the Andes mountains.

M. Harryan'a.

PHAIUS WALLICHII.

Schenley Park Conservatory.

Phaius (from *phaios*, shining; so named on account of the bright colors of the flowers). This genus contains fine, tall, terrestial plants with beautiful flowers. They grow in tropical Asia and on the islands of the Indian ocean. The leaves are plaited, the flower-stalk erect, the lip is funnel-shaped.

Phaius grandifol'ius (the finest of the species).   [s.]
P. Humboldt'tii.   [A.]
P. macran'tha (large-flowered).   [A.]
P. maculat'us (spotted).   [s.]
P. Tunkervil'lei.   [A.]
P. Wallich'ii.   [s.]
P. tuberculos'is (tubercled)   [s.]

Platyclin'is glumacea (from *platys*, broad; and *clinis*, a couch). A small but elegant Orchid with fragrant greenish white flowers. From the Philippine islands.   [A.]

Sophronit'is grandiflor'a (from *sophrona*, modest). A small Orchid with rather large scarlet flowers. From the Organ mountains.

TRIBE VANDAE.

Vanda is the Indian name. The plants of this tribe are mostly epiphytal, that is growing upon trees; many of them belong to tropical America and Asia, few to Africa. One of the distinctions between this tribe and the Epidendrae is that in the latter the cells of the anthers are separate while in the former they flow together. There are also differences in the general aspect which enables one to readily distinguish the plants of the two tribes; the two ranked leaf arrangement of many of the Vanda tribe is one of these characteristics. Darwin speaks of the Vandae as "the most remarkable of all orchids." In some genera they assume the most curious forms, resembling insects of various kinds, birds, etc. The genus *Vanda* contains a number of magnificent species, noble in form with flowers of exquisite beauty.

Vanda gigan'tea (gigantic).   [A.]
V. suavis (sweet)   [A.]
V. tricolor' planilab'ris (three-colored, flat-lipped).   [A.]
V. tricolor' Pattersoni.   [A.]

A picture of this Orchid is shown on page 80. The handsome flower, the two-ranked leaf arrangement and the aerial roots should be noticed; although an air plant it is placed in a pot for convenience; the pot contains no earth, but only moss and fragments of flower pots.

Aerides. This genus has its name from *aer*, air, in reference to the fact that the plants belonging to it are all air plants. They resemble the Vandas and have like them the leaves in two ranks.

Aer'ides crassifol'ium (thick-leaved). [A.]

A. expansum Leon'iae. [A.]

A. Fielding'ii. [A.]

A. Lob'bii. [A.]

A. odoratum (fragrant). [A.]

Angraec'um Sanderian'um, *Angurek*, the Malayan name for air plant.

A. eburn'eum. [A.]

A. sesquipet'ale. [A.]

Anguloa Clowesii, named after the Spanish naturalist Angulo. This species is found in the Andes of Columbia at a height of 5,000 to 6,000 feet.

Ada auranti'aca. The first name of this pretty Orchid was probably given as a compliment to a pretty young lady of the same name; *aurantiaca*, orange colored. A native of the Andes of Columbia, where it is found at the height of 8,500 feet. [S.]

Bras'sia verrucos'a, named after the Orchid collector Wm. Brass; *vrerucosa*, warty. A picturesque plant from Guatamala. [A. & S.]

Burlington'ia pubes'cens, named af.er the amiable and accomplished Countess of Burlington; *pubescens*, downy. [A.]

Calan'the Regnier'i, from *kalos*, beautiful and *anthos*, a flower. From Cochin China. [S.]

C. rose'a, synonym for Limatodis rosea. [S.]

C. vestit'a, clothed; from Burmah. Of this species there exist many varieties. [A.]

C. vestita var. lutea, yellow. [S.]

C. vestita var. rubra, red. [S.]

Cataset'um tridendatum. From *kata*, downward and *seta*, a bristle; referring to the two horns of the column; *tridendatum*, three-toothed. A fine Orchid with flowers four inches in diameter. [A.]

Cochliod'a sanguin'ea, from *Cochlion*, a little snail; *sanguinea*, blood red. [S.]

C. vulcan'ica. These plants used to be called

Mesospinidium, sanguineum and vulcanicum.

Coryan'thes macran'tha, from *korys*, a helmet ; and *anthos*, a flower, in reference to the shape of the lip ; *macrantha*, large flowered. A very curious but handsome Orchid.

Cyrtopod'ium punctat'um, from *kyrtos*, curved ; and *pous*, a foot, alluding to the shape of the lip ; *punctatum*, spotted. From Brazil.

Cymbid'ium eburn'eum. From *kymbe*, a boat, referring to the boat shaped lip ; *eburneum*, ivory. A large and handsome Orchid, deliciously fragrant. From the East Indies. [A. & S.]

    C. gigan'teum, from India, a large flowered species. [S.]

    C. Lowian'um, from Burmah. [S]

    C. pen'dulum, from Nepaul. [A.]

Lycaste, called after Lycaste the beautiful daughter of Priam. A genus of ornamental and mostly sweet-scented orchids. They are natives of Mexico and Central America.

    Lycas'te aromat'ica. [A.]

    L. Harrison'iæ. [A.]

    L. Skinneri, one flowered species from which many handsome varieties have been produced. [A.]

Miltonia cuneat'a, named after Viscount Milton, afterwards Earl Fitz-William. A beautiful Orchid with a wedge-shaped lip ; the flower nearly four inches in diameter. [A.]

    M. spectab'ilis moreliana. [S.]

Maxillar'ia, from *maxilla*, the jaws of an insect, which the column and lip resemble. They are natives of tropical America.

    M. aromatica. [A.]

    M. Harrisoniæ. [A.]

    M. tenuifolia, narrow-leaved. [A.]

Mor'modes pardinum, from *mormo*, a goblin, referring to the strange appearance of the flowers ; *pardinum*, panther spotted. From Mexico. Allied to Catasetum. [S.]

Odontoglossum, from *odous*, a tooth ; and *glossa*, a tongue, referring to the tooth-like processes of the lip. This genus contains over one hundred species of interesting and handsome Orchids. The flowers have spreading sepals; the lip is variously toothed and crested and its base is parallel with the long and narrow column. These Orchids inhabit the Andes of tropical America.

Odontoglos'sum citros'mum, lemon-scented.	[A.]
O. **Clowesii**.	[A.]
O. **cordat'um**, heart-shaped-lipped.	[A.]
O. **crisp'um**, curled.	[A. & S.]
O. **grand'e**.	[A.]
O. **Hal'lii**.	[A.]
O. **Ins'leayi**.	[A.]
O. **Leisden'i**.	[A.]
O. **Lindleyan'um**.	[A.]
O. **nevaden'se**.	[A.]
O. **Pescator'is**.	[A. & S.]
O. **Phalænop'sis**, moth-like.	[A.]
O. **pulchel'lum**, pretty.	[A.]
O. **Roezli**.	[A.]
O. **Rossii**.	[A.]
O. **tigrin'um**, tiger-spotted.	[A.]
O. **vexillar'ium**, standard.	[A.]

Oncidium, from *oncos*, a tumor, referring to the warty crest at the base of the lip. This genus has nearly 300 species that are known, many of the oddest shape, imitating butterflies and other insects. Some of these species are found at great height in the Andes, close to the limits of perpetual snow.

Oncid'ium albo-violaceum, white violet.	[A.]
O. **altis'simum**, highest.	[A.]
O. **amplicat'um**, broad-lipped.	[A.]
O. **Cavendishian'um**.	[A. & S.]
O. **ciliat'um**, fringe-lipped.	[A.]
O. **con'color**, one colored.	[A.]
O. **cris'pum**, curled.	[A.]
O. **cruen'tum**, bloody.	
O. **cuneat'um**, wedge-shaped-lipped.	[A.]
O. **flexuos'um**.	[A.]

O. fuscat'um. [s.]
O. Kramerian'um. [s.]
O. incur'vum. [s.]
O. Jonesian'um [A.]
O Lancean'um. [A.]
O. leopardin'um, leopard-spotted. [A.]
O. long'ipes, long stalked. [A.]
O. lur'idum, lurid. [A.]
O. Limin'gii. [A.]
O. ornithorhyn'chum, bird's bill. [A. & S.]
O. Papilio-majus, butterfly plant. [A. & S.]
O. Phalænop'sis, moth-like. [A.]
O. pulchel'lum, pretty. [A.]
O. sarcodes, flesh-like. [A.]
O. sphacelat'um, scorched. [A.]
O. splend'idum. [A.]
O. tigrin'um. [A.]
O. verrucos'um, warty. [A.]

Perister'ia elat'a, Dove flower, from *Peristera*, a dove; *elata*, tall. The column of the flower of this Orchid has two wing-like appendages, giving the center of the blossoms some resemblance to a dove. In South America it is called "Flor del Espiritu Sante." Flower of the Holy Ghost. [A. & S.]

Phalænop'sis, the Moth Orchid; from *phalaina*, a moth; and *opsis*, resemblance. Some of the most exquisite Orchids belong to this genus and some species, especially when seen at dusk, may indeed be taken for beautiful moths. They are natives of the Malayan Archipelago and Eastern India.

Phalænop'sis amab'ilis, lovely; flowers white, sometimes five inches in diameter. [A.]

Ph. grandiflor'a, large flowered. [A.]

Ph. Schillerian'a, has beautiful rose-colored flowers and spotted leaves. [A.]

Pilum'na frag'rans, Cap Orchid, from *pileus*, a cap, to which the flower has some resemblance. Also called Trichopilea candida.
[A.]

Rodriguez'ia secund'a, named after the Spanish physician and botanist Em. Rodriguez; *secunda*, side-flowering. From Trinidad. [s.]

Saccolab'ium Blum'ei, from *saccus*, a bag; and *labium*, a lip, alluding to the baggy lip. From the East Indies. [A.]

S. curvifol'ium, curved-leaved; has cinnabar-red flowers. [A.]

S. violac'ium, violet, has large showy flower clusters. [A.]

Scuticar'ia Steelii, from *scutica*, a whip, alluding to the shape of the leaves. From British Guiana. [A.]

Stanhop'ea aur'ea, named in honor of Earl Stanhope; a deliciously fragrant Orchid with golden yellow flowers. [A.]

Stanhopea Martiana, a magnificent flower. [s.]

Zygopeta'lum Mackay'i, from *zygos*, a yoke; and *petalon*, a petal. Large yellowish green flowers, spotted with purple From Brazil. [A. & s.]

TRIBE NEOTTIAE.

This tribe has its name from *neottia*, a birds nest, from the interlacing of the numerous fibrous roots; they are all terestrial and have no pseudo bulbs.

Goodyer'a dis'color, named in honor of John Goodyer a British botanist; *discolor*, two-colored. From South America. [A.]

Goodyera pubescens and repens, popularly called "Rattlesnake Plantain," are not rare in the mountains of this State.

Sobral'ea macran'tha, named after the Spanish botanist Sobral; *macrantha*, large flowered. A very large-flowered, crimson, aromatic Orchid from Mexico. [A.]

Vanilla Phalænopsis, Vanilla plant. From the Spanish *vainilla*, a little sheath, in allusion to the shape of the fruit; *Phalænopsis*, moth-like. This is a climbing Orchid from Madagascar. Tnere are about twenty species of Vanilla, the fruit of several of them is used on occount of its delicious flavor, but Vanilla plan'folia furnishes the best quality of Vanilla. [A.]

TRIBE CYPRIPEDIUM.

The name of this interesting tribe is derived from *kypris*, Venus; and *podion*, a slipper, the common name being Lady's Slipper, also Moccasson plant. The plants of this tribe are distinguished from the other Orchids by having two stamens, one on each side of the column, while the one, middle stamen, which is present in other Orchids, is here modified into a shield-like plate. The lip forms a large inflated pouch, from this the plant received its name. Bees and other insects which crawl into this pouch in search of honey, find much difficulty in making their way out again because the shield mentioned above, bars their way; in their efforts to extricate themselves they can hardly fail to get in contact with the stigma, which is situated under the shield, and if the insect has been visiting another plant of this kind before, it will be fraught with pollen and thus cause the fertilization of the stigma. In making its exit at last, the insect must squeeze itself through an opening near the stamens and becomes charged with more pollen, which may do good service in the next Orchid visited. This is only one of the many instances where the flowers exact service from the insects for providing them with sweet nourishment.

It should be mentioned yet that in the Cypripediums two sepals of the calyx are grown into one, generally standing erect or arched back of the column, while the third sepal is below the pouch-like lip. On each side of the lip is a petal, these are often narrow and long, sometimes curiously hairy or warty.

The visitor should not fail to closely examine the various forms of this peculiar tribe, of which they will find a very fine collection in the Conservatories. The following is the list.

Cyprped'ium Ar'gus.	[A.]
C. barbat'um, bearded.	[A. & S.]
C. barbatum nigrum.	[A.]
C. Boxal'lii.	[A. & S.]
C. caricin'um, sedge-like.	[A.]
C. caudat'um, tailed.	[A. & S.]
C. Chamberlainian'um.	[A. & S.]
C. Dayan'um.	[S.]
C. Dominian'um.	[A.]
C. Harrisian'um.	[A. & S.]

C. Haynaldian'um. [S.]
C. hirsutis'simum, very hairy. [A. & S.]
C. insig'ne, remarkable. [A. & S.]
C. Japon'icum, Japanese. [S.]
C. Lawrencian'um (see picture). [A. & S.]
C. laevigat'um, smooth. [S.]
C. longifol'ium, long-leaved. [A.]
C. Parish'ium. [S.]
C. Roez'lii. [A.]
C. Seden'ii. [A.]
C. Spicerian'um. [A. & S.]
C. venus'tum, beautiful. [A. & S.]
C. villosum. [A. & S.]

Nearly all of these species are from India, the Malayan Archipelago or the Philippine Islands.

CYPRIPEDIUM LAWRENCIANUM.
Schenley Park Conservatory.

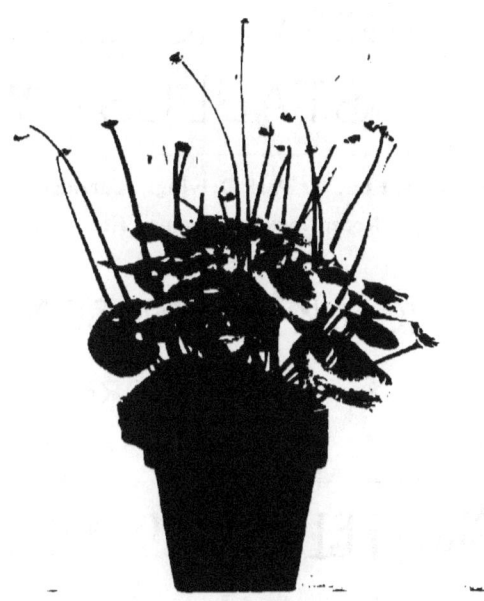

CYCLAMEN PERSICUM.
(Persian Violet.)
Schenley Park Conservatory.
(See page 105)

L. of C.

SCHENLEY PARK
LIVERY, BOARDING AND SALES STABLES.

All Kinds of Traps, Etc., for Park Driving. Carriages for Weddings, Receptions, Shopping, Etc.

VICTOR G. WILSON,
3994 FORBES STREET.

TELEPHONE 4016. NEAR PARK ENTRANCE.

A LONG FELT NEED SUPPLIED.

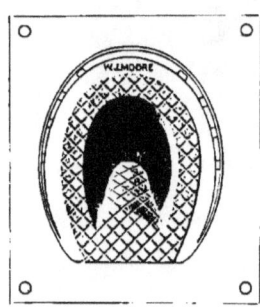

Have You Tried?

The Columbian Noiseless Rubber Pad, for the cure of sore feet, corns and all ailments caused by the hard streets. These Pads prevent a horse from slipping and are the only safe way for shoeing high knee acting horses.

◇For Sale and Put On By◇ **W. J. MOORE,**
OAKLAND AVENUE,

Agent for Spelterine Hoof Stuffing.

Pittsburgh, Pa.

. . . TELEPHONE 4096. . . .

THE FLOWERS.

THE FLOWERS.

The Flowers.

Nothing has contributed so much to make the Conservatories popular as the Floral Exhibits varying from month to month. Mr. Hamilton's Chrysanthemum shows at Allegheny Conservatory have long enjoyed the reputation of being among the finest in the country; but Schenley Park Conservatory with its two spacious wings devoted to exhibitions, has made a floral display possible, which in tasteful arrangement of details and grand totality is simply magnificent.

To see a beautiful flower is a thing of pleasure, but to see thousands of them massed together, filling a large house with glowing color and often also sweet fragrance, makes upon the beholder an almost overwhelming impression. And such floral feasts are now in store for the visitor of the Conservatory every month in the year.

Although it may be said that in a greenhouse there is perpetual summer, the flowers have their regular seasons for blooming, somewhat in keeping with the blossoming time in the native home of the plants; taking advantage of this fact it is possible for the managers to present to the public gaze dissolving views of ever varying splendor—to weave a floral wreath around the year.

Superintendent Bennett has outlined a flower calendar which will indicate what will be the prominent features in the exhibition houses from season to season; the programme is certainly full of interest and promise and should induce flower lovers to pay visits to the Conservatory at regular intervals, so as not to miss a single part of this delightful procession of Flora's loveliest children.

CALENDAR OF FLORAL EXHIBITS.

January will be greeted by the CYCLAMENS in their fullest glory; the Cyclamen exhibit last winter was a revelation to many a visitor; few of them had ever seen these flowers as grand in size and as varied in color, and probably none had ever seen them in such numbers. PRIMROSES will not stay long behind, some fine ORCHIDS will be in flower and a gorgeous array of TULIPS and NARCISSUS will soon follow, lasting well into the month of

February, when the CINERARIAS will make their grand debut, as they did this year. GERANIUMS, too, in variety of size and color seldom seen before, may plead a chance to be admired. IXIAS and SPARAXIS will lend further variety, while lovely FREESIAS will take care that sweet fragrance is not missing.

March, is the time for EASTER LILIES and AZALIAS; whoever saw this elegant combination this year will not want to miss them in any future year. HYACINTHS will be there too; Genistas will mingle their gold with the masses of white and pink and red, enhancing those colors as a golden frame often does the pigments on an artists canvas.

April will see AMARYLLIS in bloom and LILIES of various kinds arrayed more gorgeously than Solomon in all his glory. These will probably fill one house while the other will be one immense bed of CALCEOLARIAS, those odd-shaped flowers of endless varieties of color. They will continue to bloom far into the month of

May until crowded out by RHODODENDRONS and HYDRANGIAS, both characterized by immense flower clusters, the ones of gorgeous tints like the Azalias, the others more delicate, also more lasting.

June, the month of Roses, must be devoted mainly to this queen of flowers which will never lose its place in the hearts of men. Other displays may be more brilliant, more grand or more curious, but none will awaken more genuine admiration, more heartfelt joy than the sight of the beautiful roses. This flower seems more closely interwoven with our tender emotions than any other.

> Which flower has been honored most
> In song, in rhyme or prose?
> By poets, sages, lovers—say?
> Ah, 'tis the lovely Rose!

In the Aquatic department the WATER LILIES will be in their fullest glory.

July will see the elegant GLOXINIAS at their best; the BEGONIAS will show that not only their foliage, but also their flowers are to be admired.

August may surprise us with a glorious array of FUCHSIAS, their pendulous flowers glowing in all the colors from pure white to darkest purple. Tender and lovely ACHIMENES will make friends among many visitors who did not know the flower well enough to appreciate it as it deserves to be. CALADIUMS will form a proud and splendid background. These may last well into

September, when TYD.EAS in their many hybrid forms and GESNERAS will also make their appearance.

October will be "gala" month for both Conservatories; the CHRYSANTHEMUM show, an annual event among the florists and flower loving public in nearly every large city, will become more interesting than ever on account of the rivalry between the two Conservatories and our enterprising florists, who have for years paid much attention to the cultivation of the newest and best varieties and with excellent success. This exhibit will probably begin about the middle of October and last till the middle of

November. Another set of flowering BEGONIAS will be on hand by this time and various other plants in bloom will help to fill the show houses.

December will gradually lead the way to the new year, the CYCLAMENS, being ready again, the PRIMROSES starting, ORCHIDS beginning to bloom freely and other plants which have been mentioned under January.

Of course the changing flowers do not keep exact time with the months. The different kinds as they appear in their rotation "dovetail" into each other; nor are those named above all that may be expected; new and charming species will be added from time to time each one a new surprise to the hosts of visitors.

A few remarks about each of the flowers mentioned may be of interest to the reader.

Cyc'lamen persicum, the Persian Violet. From *kyklos*, circular, referring to the spiral peduncle or probably to the round tuber. These beautiful plants are natives of Persia, they belong to the primrose family and should be, more appropriately called "Persian Primrose." Several smaller species grow quite abundantly in the mountainous parts of Europe; they produce, close to the surface of the ground, bulb-like tubers which the pigs root up and munch with great satisfaction, hence the unpoetic name of "sow-bread," under which the plant is known in Europe. The Persian Cyclamen, the largest of the natural species, has been wonderfully improved by skillful cultivation, so that they are produced now remarkably large in size and many color varieties from pure white through all shades of pink to the deepest glowing crimson. The finest specimens are produced from the seed; after the plant has bloomed once, it generally deteriorates and the next season the flowers are found to be smaller. See illustration on opposite page.

Prim'ula, Primrose. From *primus*, first, being among the first flowers that bloom in the spring. The English Cowslip (Primula officinalis) and Oxlip (Primula elatior) are primroses; the Germans called them " Schluesselblumen," meaning "key-flowers," because they open the gates of spring.

Primula sinen'sis, the Chinese Primrose, is the one chiefly cultivated by florists on account of its profusion of flowers and its endless color varieties; it is well adapted for such floral exhibits as are prepared in the Conservatories. But even a single plant is an object of beauty and a most graceful adornment of the flower stand at home.

Tul'ipa, Tulip. The name is said to be derived from the Turkish, *tulband*, a turban. The Tulips belong to the Lily family and are natives of southern Europe, northern Africa, western and central Asia. There are many species and hundreds, probably thousands of varieties have been produced during the centuries that this plant has been cultivated. In the seventeenth century Tulips were an object of trade and of speculation and gambling, such as grain and oil are to-day. New varieties are produced from seeds taken from plants which have been hybridized. Hybridization is a process long ago known and practiced and to it we owe the color varieties of many cultivated plants. The following will explain the process: Suppose a white tulip is deprived of its stamens before they ripen their pollen and the pollen from the stamen of a red tulip is dusted upon the stigma, thus fertilizing the ovules (the young seeds) of the white flower by means of the stamens of the red. The seeds, when matured and planted, will in all probability produce a flower with a mixture of white and red in their petals. The variety thus obtained is generally multiplied by means of the offsets which the bulb produces, for the seed of the variety will as a rule produce plants with a tendency to return to the simpler, one colored form.

Narcissus. Who does not know the story of this youth who, in love with himself, stood at the water's edge to see his image in the well; but as such love has little comfort in it, he wasted away until he was reduced to a flower, while lovely Echo, the poor nymph who loved him, pined away until nothing was left of her, but her voice, which still resides in the woods and which you have heard there, I suppose. The Narcissus plant is classed among the Amaryllis family, it inhabits the same countries as the Tulip and like that genus is produced in innumerable varieties. One of the best known species is

Narcis'sus poet'icus, the Poet's flower, which grows wild in Europe;

Narcissus Jonquil'la, one of the Jonquil's, and

Narcissus pseudo-Narcissus, one of the Daffodils, may be mentioned as a characteristic representative of this large and handsome host of flowers which my fair readers love so well because they are charming harbingers of spring.

Cineraria cruenta, from *cinerea*, ash colored; the name refers to the ashen appearance of the under side of the leaves; *cruenta*, purple-leaved. From this species, which is a native of the Canary Islands all the wonderful varieties exhibited in the Conservatories are derived; it is a most wonderful plant for producing color varieties which range through all shades of blue to pure white, also merge into crimson and violet. The Cinerarias belong to the Composite or Sunflower family; thus, what appears to be one flower to the uninitiated is a whole cluster of flowers, the petal-like parts being ray flowers and those generally mistaken for stamens being disk flowers. German florists have done much to bring these plants to the present perfection. (See illustration on page 101.)

Geranium, Crane's Bill; from *geranion*, the old Greek name used by Dioscorides and derived from *geranos*, a crane, which refers like the English name to the beak-like fruit of the plant. What could be said here about Geraniums that the reader does not know already? That they belong to the most grateful flowers to grow in garden and house is certain; they are so easy to raise and they brighten up the surroundings by their cheerful colors. What may be new to some is that most of our garden Geraniums are not Geraniums in the eyes of the botanists; they call them Pelargonium, or Stork's Bill; the principal difference between the two can be found in the shape of the petals; in the Geraniums the corolla is regular, that is, the petals are alike in size and shape; in the Pelargonias the flowers are irregular, some petals being larger than the others or differently marked. Geraniums are found wild in this country and in Europe, the handsomer species are mostly from Africa and the Canary Islands. Our own Geranium maculatum (spotted), so plentifully found in the woods around Pittsburg is a rather handsome plant. The Pelargoniums are for the most part natives of the Cape of Good Hope, and it is a collection of these, in many varieties—velvety, glossy, hairy, dissected, ivy, fragrant-leaved, small and large-flowered, of every hue and tint between white and scarlet, to deep purple, that is promised for next spring.

Pelargon'ium in'quinans, stained flowered, is the parent of most of the so called " Scarlet Geraniums " of our own gardens.

Pelargonium zonal'e, the " Horseshoe Geranium " is also the parent of many cultivated varieties which vary from scarlet and crimson through all shades of red to pure white. The leaves have generally a dark mark resembling a horseshoe.

Ixia; this Greek name is used by Theophrastus for bird-lime and refers to the sticky juice of these plants; they belong to the Iris family and are bulbous plants coming from South Africa. There are many pretty varieties, pink, flesh-colored, orange, yellowish-white to pure white, also striped and variegated. Ixia odorata is very fragrant.

Sparaxis, from *sparasso*, to tear, so named on account of the lacerated appearance of the spathe which encloses the flower when in the bud. These exquisite little plants are closely related to Ixia, and are from the Cape of Good Hope.

Freesia, the name is of unknown origin; these pretty bulbous plants with grass-like leaves and white, exceedingly fragrant flowers are becoming more and more popular. They also belong to the Iris family. The species we meet in the Conservatories are:

Freesia refracta, bent back, and

Freesia refracta alba, pure white. They are all from the Cape of Good Hope.

Azalea, the name of this well known genus is derived from *azaleos,* which means dry, arid, alluding to the nature of the region in which some of the finest species grow. The Azaleas belong to the Heath family which furnishes a number of beautiful and also some useful plants. The Mountain Laurel, the Sheep Laurel, the Arbutus, the Heather of Scotland and Ireland, the Alpenrose of the Alps, the Ericas and Epacris of Africa and others are prized for their flowers, the Huckleberry and Cranberry are valued for their fruit. The Azaleas used to be classed with the Rhododendrons, but there seems sufficient reason for placing them in different genera; the Azaleas have generally five stamens while the Rhododendrons have ten; the Azaleas bloom at an earlier season.

There are few shrubs more popular in Pittsburg than Azaleas; being generally hardy, of easy culture and blooming early and abundantly in Spring when flowers are most appreciated, they are planted by almost every fortunate owner of a garden where there is room for some shrubbery.

Florists divide the Azaleas into two groups, one of them includes the Ghent or American Azaleas; to these belong most of the hardy plants found in the gardens, several of them are from stock native in this country; such as AZALEA ARBORESCENS, tree-like; A. CALENDULACEA, marigold-like, with red and orange flowers; A. NUDIFLORA, naked flowered, known under the name of Mountain Honeysuckle; A. SPECIOSA, showy; A. VISCOSA, clammy, and others are species found in the Allegheny mountains, but most of them have been greatly improved by the Florists of Ghent and other parts of Belgium. The other group are the Indian or Chinese Azaleas; these are evergreen varieties obtained from Azalia indica, which are cultivated in greenhouses and have a longer season for blooming. Other species of this group are:

Azalea amœna, pleasing, and A. SINENSIS, Chinese. What rich varieties have been produced from these everyone knows who has been visiting Schenley Park Conservatory last March or Allegheny Conservatory in April.

Rhododendrons shall be discussed right here because they are so closely related to the Azaleas. The meaning of the name is "Rose Tree" and a common English name for it is Rose-Bay. Rhododendrons are found in Europe, Asia and America, our own country possessing some of the finest species of hardy plants such as RHODODENDRON MAXIMUM, the American Great Laurel, of our Alleghenies; R. CATAWBIENSE, the Southern Laurel and R. CALIFORNICUM the California Laurel. German florists have taken hold of these and others and by hybridizing and other means, have produced a great number of charming varieties, such as filled the exhibition house of Schenley Park last May. Thus we import our own Azalias from Belgium and other Rhododendrons from Germany. But the American florists are not sleeping and the time probably is not far distant when they will lead the world in the cultivation of the native species. Among the foreign Rhododendrons, of which there are many species, the most prominent are RHODODENDRON NUTTALLII, from India; R. ARBOREUM, tree-like from the Himalayas; R. AUCKLANDII; R. EDGEWORTHII, from Sikkim;

R. JASMINIFLORUM, jasmine-flowered, from Malacca ; R. JAVANICUM; from Java; R. PONTICUM, from Asia Minor, is the species from which the most showy hybrids are raised.

THE LILIES.

Consider them, how they bloom ! There is no more proud, no more majestic flower ; the petals resplendent white or glowing red, or brilliantly striped and spotted and of the richest fragrance which is sometimes almost intoxicating. The first Lily of the season and certainly one of the choicest is

Lillium Harrisii, the Easter Lily, because it expands its glorious white flowers about Easter time as if intended by nature to assist in the celebration of the resurrection of the Lord. It is also called Bermuda Lily because it is cultivated in those islands in enormous numbers ; though the plant is said to be originally a Japanese species. It is probably a variety of Lillium longiflorum the long-flowered, choice Lily, but it has marked characteristics of its own and is much easier to propagate.

To the Lily family belong also the following :

Hyacin'thus. Hyacinth, the ancient name used by Homer for the Iris. These well known and well loved flowers need no introduction ; they precede the Lilies and are among the first flowers that venture out after the winter. The numerous varieties are derived from

Hyacinthus orientalis, the Eastern Hyacinth, which is a native of Syria, and the

H. orientalis provencialis, the Provence Hyacinth which is found in Southern France, Switzerland and Italy.

Agapan'thus umbellat'us, the African Lily ; from *agape*, to love ; *anthos*, a flower ; *umbellatus*, the flowers forming an umbel. This and several other liliaceous flowers mentioned below can be found in Schenley Park Palm House ; it will attract attention by its tropical aspect, having long narrow, but thickish leaves and a still longer flower stalk crowned with a large umbel of beautiful blue or white (variety alba) flowers.

Anther'icum picturat'um, *anthos*, a flower ; *kerkos*, a hedge, in reference to the tall flower stem. This and the two following are from the Cape of Good Hope ; tall Lilies with Grass like leaves and slender flower-stalks.

A. vittat'ta, spotted.

A. variegat'a, variegated ; the leaves being striped with white.

Aspidis'tria lurida, *aspidion,* a little round shield ; probably so named on account of the little mushroom shaped stigma of the flower. From China.

A. variegata.

The last five are in the Allegheny Conservatory.

AMARYLLIS FAMILY.

Amaryl'lis. This flower is the namesake of Amaryllis, the country woman, mentioned by Theocritus and Virgil. With the handsome, large, lily-like flowers the reader is, no doubt, well acquainted. The principal species is

A. Belladonna, the Belladonna Lily, which comes from the West Indies and from which several varieties have been produced ; the pale variety is

A. Belladonna pallida.

Many hybrids which figure in florists catalogues as Amaryllis belong to the genus Hippeastrum,

Hippeas'trum reticul'atum, the Knight's Star. *Hippus,* a knight ; *astron,* a star ; *reticulatum,* netted.

Pancrat'ium, *pan,* all ; *kratys,* potent ; it was once believed to have powerful medicinal purposes. These lily-like flowers are white and fragrant and often have long and narrow perianth lobes which give them a peculiar appearance; they are from the Mediterranean region, the Canary Islands and the West Indies.

Imantophyl'lum miniat'um. *Imas,* a leather thong; *phyllon,* a leaf; *miniatum,* brick colored. This plant is distinguished by its thickish two-ranked leaves from which rise a fleshy flower-stalk bearing an umbel of very large, showy flowers of a bright orange-tinted vermillion.

Doryan'thes Pal'meri, *dori,* a spear ; *anthos,* a flower ; the long flower-stem having been compared to the shaft of a spear. A very handsome plant of the Amaryllis order, having narrow, stiff and pointed leaves six feet long, growing in a tuft from the midst of which the long scape arises, bearing a large spike of red flowers. From Queensland.

Euc'haris amazon'ica, also named

E. grandiflora, from *eu*, well; and *charis*, grace; so named on account of its exceeding gracefulness. The flower resembles somewhat a Narcissus, but it is much larger, sometimes four inches in width. From New Grenada.

Crin'um amab'ile, from *krinon*, the Greek word for Lily; *amabile*, lovely. Another beautiful species having bright red, fragrant flowers of which sometimes twenty to thirty grow in one umbel. From Sumatra.

These compose the Amaryllis collection in Allegheny Conservatory. The Agaves and some other plants of the same order have been mentioned in a former chapter.

Genista, this is the old Latin name for the plant; from it the Plantagenets took their name having chosen "Planta Genista" as their emblem. This genus is distributed all over Europe, western Asia and northern Africa. The flowers are all yellow and butterfly-shaped, the plant belonging to the Pulse family to which the Pea, Bean, Locust and many other well known plants belong. Genista tinctoria, the Dyer's Greenweed, used to be largely cultivated for the preparation of a dye known under the name of Kendall Green.

Cyt'isus canarien'sis, from the Canary islands, is the plant generally called Genista canariensis.

Calceolar'ia, Slipperwort or Fisherman's Basket; the Latin name is derived from *calceolus*, a little slipper; according to some authorities, however, the name has been chosen in honor of F. Calceolari, an Italian botanist of the sixteenth century. The plant belongs to the Figwort family (Scrophulariaceae), to which the Mullen and Butter-and-Eggs of our meadows belong. The Calceolarias are South American plants, coming principally from Chile and Peru, but the beautiful hybrids which have been so much admired in the Conservatories are mostly the product of the English florists art. There are shrubby and herbaceous Calceolarias, the latter being more commonly cultivated in our greenhouses. The innumerable hybrids are derived principally from

Calceolaria amplexicaul'is, leaf clasping the stem.
C. arachnoid'ea, cobwebby.
C. corymbos'a flower-cluster being level.
C. integrifol'ia, entire leaved.
C. purpur'ea, purple.
C. thyrsiflor'a, flower-cluster pyramid-like.

CALCEOLARIA,
(Fisherman's Basket.)
Schenley Park Conservatory.

Those who have visited the exquisite Calceolaria exhibit in Schenley Park Conservatory or the annual display of these flowers in Allegheny, need not be told what an odd freak of nature these flowers represent; the corolla has a very short tube and two lips of which the upper one is rather small, the lower very large and baggy somewhat resembling a fisherman's creel; the two stamens and the pistil are hidden in the upper lip. As there is nothing in nature without a design, or use, or reason, the question is a legitimate one: What is the purpose of this arrangement?

The author has not found yet a satisfactory answer. As the lower lip has much resemblance to the baggy lip of the Cypripedium, a similar purpose suggests itself; and it may be that the arrangement is a contrivance to entrap insects to insure cross-fertilization. The matter is still open for speculation and investigation.

The most important feature which makes the Calceolarias so valuable to the florists is their susceptibility to hybridization and variation; among thousands of plants exhibited there are often not two exactly alike in color and markings. The predominant color is yellow, but this ranges from the most delicate cream to the richest brown and red; the markings may be simple dots, leopard spots, crescents, rings or innumerable other designs; others are one colored or shaded from light to dark.

Hydrangea, also called Hortensias; the name is derived from *hydor* water; and *aggeios*, a vessel; the cup-shaped fruit having been compared to a water vase. These plants belong to the Sanifrage family and are related to the Goosberry, Currant and the Mock Orange of our gardens. The home of most of the cultivated species is China, Japan, the Himalaya mountains and Java, but our own native

Hydrangia arbores'cens, tree-like, is also a rather handsome shrub and can, no doubt, be much improved by cultivation.

H. hortensis, is the common garden Hydrangea.

H. hortensis japon'ica, the Japanese Hydrangea, is the one with blue and white flowers.

H. paniculat'a grandiflora, large flowered, panicled, is the handsome shrubby Hydrangea which blooms late in summer and in autumn.

ROSES.

There is a time in the destiny of every properly conditioned young man when he begins to cultivate the language of flowers; much to the satisfaction of some young lady with estimable qualities. Now for the the enlightenment of that young man be it stated, that, to find the most eloquent words for his case in the vocabulary of said language, he must turn to the chapter on Roses. He will find there the interpreters for all the forms and moods of love. For pure and holy love the white rose is the matchless symbol, while the red portrays the burning passion, the yellow rose the pangs of jealousy, and for all gradations there are found the appropriate shades. And while the young man is trying to select what color will be most befitting his condition, he must not forget that the odor, too, is an important factor. There is the beautiful but odorless, the faintly, the delicately fragrant, the deliciously, the exquisitely odorous to choose from. But mark, young man, among all the lovely, the glowing, fiery roses, there is no spotted rose and among all the various odors of this flower there is none intoxicating. From this fact draw your own moral.

Rose fanciers have been trying long and hard to produce a black rose, but they find it difficult; this flower does not take kindly to the sombre hues. The author saw the nearest approach to a black rose at a Rose Exhibition in Hamburg; the flower was placed in the center of a wreath of fairest roses and looked dark indeed, still there seemed a secret fire in its petals and withal it appeared like a production of black art, uncanny, yet fascinating. The roses in their natural state have five petals and many stamens. The cultivated roses get their many petals at the expense of the stamens and in some specimens no stamens at all are left; frequently stamens can be found in the transition state, being half petal, half stamen; sometimes the calyx is found to consist of fully developed leaves instead of the ordinary sepals; these and other peculiarities observed in the rose and other flowers, reveal an important fact which is now well established and it is this: Flowers are altered; branches modified for the purpose of producing fruit and seed. An ordinary branch consists of stem and leaves; in the flower-branch the leaves are modified, forming sepals, petals, stamens, pistils, and they are placed in close whorls at the end of the stem (peduncle). Occasionally some of these parts revert to the simple form from which they were evolved, thus betraying their origin.

The genus ROSA, the old Latin name for the flower, belongs to the Rose family (Rosaceae), to which nearly all our fruit trees belong.

But let us come to the rose exhibit. For purposes of making a large display, the dwarf roses, which are budded quite low near the ground, are the most convenient; high stemmed roses are very fine but much harder to manage in a greenhouse. The roses which are so much admired at the Schenley Park Conservatory, just at this writing (June 1894), are all imported from Holland; the list is given below and contains many well known varieties, but also many new forms seen for the first time in Pittsburg; next year, after having been transplanted into larger pots and having accommodated themselves to the new surroundings, the Roses will make even a grander show than this year. These exhibits should be made a study; among the new varieties are many well marked with new characteristics and well worth close examination. You may make new and lasting friends among them.

It would be a difficult task to find out the pedigrees of each of these varieties; they are the result of the care and cunning of several generations of florists who make Rose "breeding" a specialty. Here is the list.

DWARF ROSES.

Magna Charta, rosy carmine.
Ulrich Brunner, cherry red, very large.
General Jacqueminot, crimson.
Abel Carriere, dark crimson.
Mme. Gabriel Luizet, delicate pale rose.
Prince Camille de Rohan, velvety crimson.
Gloire de Dijon, fawn with salmon.
Mrs. John Laing, brilliant rosy pink.
Baroness Rothschild, pale flesh.
Pæonia, carmine.
Gloire de Margottin, deep velvety crimson.
John Veitch, vermilion.
Elizabeth Vigneron, dark rose.
Pius IX, cherry red.
Perle Blanche, white.
Anna de Diesbach, brilliant rosy pink
Geant des Battailes, vivid scarlet crimson.

Captain Christy, delicate rose.
John Hopper, rose with crimson center.
General Washington, crimson.
Alfred Colomb, bright carmine.
La France, lilac rose.
Duke of Edinburgh, vermillion.
Gloire Lyona'se, yellowish white.
Paul Neyron, dark rose, very large.
Duke of Teck, brilliant scarlet carmine.
Marie Bauman, bright carmine.
Charles Lamb, clear fiery red.
Mabel Morrison, pure white.
Heinrich Schultheis, rose.
Auguste Mie, light glossy pink.
Madame George Bruant, white.
Fisher Holmes, velvety crimson.
Merveille de Lyon, white.
Marshall P. Wilder, bright carmine.
Miss Bosanquet, delicate flesh.
Zepherine Drouhin, carmine with purple.
Countess of Oxford, bright carmine.
Gloire Bourg de la Reine, scarlet red.
Empress, white.
Duke of Montpensier, carmine.
Monsieur Boncenne, dark velvety crimson.
Hippolyte Jamain, bright rose with carmine.
Reine Marie de Henriette, crimson, fine bud.
Chestnut Hybrid, crimson, fine climbers.
Prince of Wales, silvery white.
Lord Bacon, brilliant scarlet crimson.
Mme. Victor Verdier, cherry red.
Mme. Alfred Carrier, light rose, changing to white.
Lady Mary Fits-William, delicate rose.
White Baroness, pure white.
Baron de Bonstettin, dark crimson.
Princess Louise Victoria, delicate rose.
Compte de Paris, Crimson.
Jean Liabaud, dark crimson.
Duke of Connaught, velvety crimson.
Triomphe d'Angers, vivid purple with violet.
Dr. Baillon, crimson.
Elsie Bœlle, white, tinged with rose.

Gloxinia. Professor Benjamin Peter Gloxin, a botanist of Colmar in Germany, has been greatly honored by his name being bestowed upon one of the most beautiful of garden flowers. The Gloxinias belong to the Gesnerwort family (Gesneraceae) and are all natives of tropical America. The beauty of the large, well shaped corolla richly and variously colored, is enhanced by the background of a rich velvety foliage. Exhibited in large masses they present a glorious picture, the large range of colors and the fine color combinations in single flowers being especially noticeable. To the botanist most of these cultivated varieties are known as Sinningias, named after Wilhelm Sinning, gardener to the University of Bonn, and

Sinnin'gia specios'a, showy, is the species from which most of the beautiful varieties, known as Gloxinias, are derived from. Examine them closely, they deserve it. See whether among the many hundreds of specimens you can find two alike in combinations and markings; observe also how prettily the stamens stick their heads together. Other plants of the same family may be mentioned right here, they will appear soon after the Gloxinias.

Ges'nera, both the family and this genus have been named after the celebrated botanist Conrad Gesner of Zurich; most of these plants are natives of Brazil, a few are from the northern part of South America. The plants grow from tubers; the flowers have long, tubular corollas, and are of rich scarlet, vermillion, orange or yellow colors.

Achimen'es belong to the same family; the name comes from *cheimaino*, to suffer from cold, indicating that these plants are very tender; their home is Mexico, Central America and the West Indies; these plants have peculiar, scaly rootstalks and handsome flowers, the corolla forming a long, narrow tube opening into a wide and showy border. The varieties are too numerous to mention; the most prominent species are

Achimenes coccin'ea, scarlet flowered and

A. longiflor'a, long-flowered, which has violet flowers.

Tydæa, named after *Tydeus*, a son of Œneus, king of Calydon, is a genus much resembling the one just discussed and the plants belonging to it are more generally known as Achimenes; they are natives of tropical America; probably the most distinct species is

Tydæa amabilis, lovely; a plant with hairy leaves and flowers, rose colored, with purplish dots.

Fuch'sias, named after the German botanist of the sixteenth century, Leonard Fuchs; to pronounce this plant *Fook'sia* would therefore be more correct than *Fu'shia*. This exceedingly well known and popular genus belongs to the Evening Primrose family (Onagraceae); its home is Mexico, Central America and the western coast of South America, where this plant has a shrubby and tree-like character. The species most commonly cultivated is

Fuchsia macrostem'a, (having large stamens) in its several, distinct varieties, GLOBOSA (globose), CONICA (conical), GRACILIS (graceful), PUMILA (dwarf) and RICCARTONI; these are hardy Fuchsias; among the species that will do well only with inside culture are

F. **bolivian'a,** from Bolivia.

F. **ful'gens,** glowing, with scarlet flowers, two inches long.

F. **corymbiflor'a,** flowers forming a corymb, that is standing all about the same level; the flowers are long and narrow and form a dense cluster.

F. **cordifol'ia,** heart-shape-leaved.

F. **serratifol'ia,** saw-edge-leaved.

F. **microphyl'la,** small-leaved.

F. **procum'bens,** creeping; this is one of the two species that have been found in New Zealand.

F. **splend'ens,** splendid, a very showy species. From these and a few other species all the varieties have been produced which are known under many fancy names and which are the delight of millions of flower-loving people and the joy of the florist who finds a ready sale for these flowers.

The story goes that the first Fuchsia was discovered by Father Plumer and dedicated to Leonard Fuchs about 200 years ago, but its introduction into England dates only 100 years back, when a sailor brought one of the plants to his wife at Hammersmith; Mr. James Lee, a nurseryman, saw it in the window and recognizing at once a novel and beautiful flower, he bought it, not without some difficulties, carefully raised it, found it easy to propagate it from slips and in a few years was able to sell Fuchsias in large numbers.

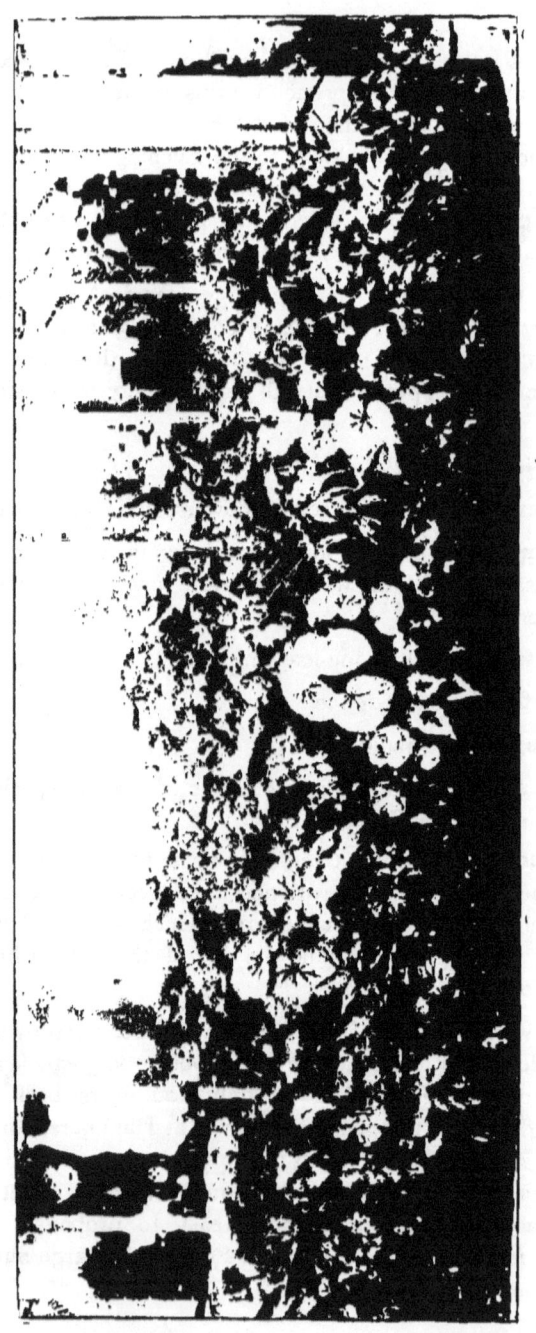
Group of Begonias.
Schenley Park Conservatory.

BEGONIAS.

The Begonias cannot be dismissed with a few remarks; a large book would have to be written to name each variety and to give adequate praise to all those who deserve it; but this book is already getting larger than the author intended; therefore the Begonias must excuse if only a short article is devoted to them. They may recognize the author's high appreciation in the fact that he has devoted to them three full page illustrations, which are works of art as every one will concede. They are wonderful plants, the Begonias, and Monsieur Michael Begon, the Frenchman who not only loved flowers, but patronized florists and encouraged botanists, has been honored more than he imagined—nay, he has been immortalized—by having his name bestowed upon this genus and upon the family (Begoniaceae) to which it belongs. Is there any other plant so characteristic in appearance and so profuse in variations; not in variations of flowers only, and these are wonderful enough, but even more so in variations of the leaf. Though about this later on.

The Begonias are found growing wild in all tropical moist countries, especially South America and India. The flowers are often showy and their colors range from white to deep red. There is no distinction between calyx and corolla; the parts of the floral envelope (perianth) being all petal like and numbering from two to five. Staminate and pistillate flowers are separate, but grow on the same plant; the stamens are generally many in a bundle and often unite below in a tube; the pistillate flowers are readily recognized by their three-winged ovary and the three stigmas. The leaves are peculiarly unequal sided and have been compared with elephant's ears.

Horticulturists divide the Begonias into tuberous-rooted, and shrubby. The tuberous rooted Begonias have been introduced from the Andes mountains, and from them, by means of careful selection and cross fertilization, many new varieties with magnificent flowers have been produced; flowers measuring from four to six inches across not being rare.

The types from which most of the varieties of Begonias with tuberous root stalks are derived from are BEGONIA VEITCHII, B. ROSAEFLORA and B. BOLIVIENSII.

BEGONIA LEAVES. Plate No. 1.

BEGONIA LEAVES.

(Shown on opposite page.)

PLATE. I.

1. Begonia Rex.
2. Alice White.
3. Anna Dorner.
4. Pauline Rothschild.
5. Argentea guttata, silver spotted.
6. Mrs. Shepherd.
7. Madame Lionett.
8. Mirabunda.
9. Madame Leboucq.
10. Caroliniaefolia, one of the few species with compound leaves; from Mexico.
11. Albo picta, white painted; a species from Brazil.

The following is a list of the SOUTH AMERICAN and MEXICAN Begonias, that is, the TUBEROUS ROOTED species and their hybrids, which the author found in the Conservatories.

Begonia albo-picta, white painted.	[S.]
B. argyrostigma picta, silver-spotted, painted.	[A. & S.]
B. carolinæfolia.	[A.]
B. coccinea, red.	[S.]
B. compta.	[S.]
B. conchaefolia.	[S.]
B. Gilsoni.	[A.]
B. glauca, covered with bloom.	[S.]
B. glaucophylla scandens, glaucous-leaved, climbing.	[S.]
B. incarnata superba, fleshy, superb.	[A.]
B. incarnata metallica, metallic.	[A.]
B. Ingramii.	[S.]
B. manicata, tunicated.	[A. & S.]
B. odorata, fragrant.	[A.]
B. olbia, rich.	[A.]
B. ricinifolia, Castor-Bean-leaved.	[S.]
B. rubra.	
B. sanguinea, bloody.	[A.]
B. scandens, climbing	[S.]
B. sceptrum, princely.	[S.]
B. Schmidtiana.	[A. & S]

B. semperflorens rosea. [s.]
B. semperflorens gigantea. [A. & S.]
B. Soundersi. [s.]
B. Standishii. [s.]
B. sub-peltata nigricans, nearly shield-shaped, dark. [s.]
B. Thurstonii. [s.]
B. Vernon. [s.]
B. Weltoniensis alba. [A. & S]

The so-called shrubby Begonias are those which excel principally in the size, form and wonderful design and color variations of the leaves. Whoever observes the Begonias, so profusely introduced among the rock-work in Schenley Park Palm House, will concede that for decorative gardening they are simply unmatched. Let me advise you, fair reader, at your next visit to Schenley Park or Allegheny Conservatory, to step into the department for tropical plants and to examine closely the Begonia leaves; if you do not draw inspiration from the infinite variety of design and exquisiteness of color blending, I shall be much disappointed.

Begonia Rex, the Royal Begonia, introduced from Assam, is the species from which most of the ornamental-leaved varieties are derived. These varieties are legion; most of them are named after some personage, famous in history, prominent in society or dear to the heart of the gardener who was so fortunate in producing a new variety worthy of being perpetuated. It is certainly an enviable, but well deserved privilege of the horticulturist to be permitted and able to immortalize his friends by giving his or her name to a beautiful plant which will be introduced into and admired by all the civilized world and will be handed down to future generations. The following is a list of Begonias from CHINA, JAPAN and other parts of Asia and the East Indies; they are the SHRUBBY Begonias, most of them being derived from

Begonia rex, the king.
Alice White.
Anna Dorner.   Bettina Rothschild.
Clementina.   Comptess Erdody.
Diadema.   Dorothy.
Duke of Veragua.   Elegant.
Evansiana.   Feastii.
Flora Hill.   Greyhound.
Indiana.   Isabella Bellon.

Lady Slade. La France.
Llewellyn. Louis Closson.
Madame Leboucq. Madame Lionnet.
Madame Luizet. Madame Montet.
Madame Treyve. Marguerite.
McBethy. M. de Lesseps.
M. Paraert. M. E. W. Scripps.
Mrs. Shepherd. Mirabunda.
Nickle Plate. Pauline Rothschild.
Perle Hunfeldt. President Carnot.
Richmond Beauty. Smaragdina.
Sterling. Souvenir de Joseph Main.
Sutherlandi. Ville de Neully.
Whittier. Wyoming.

These can all be found in Schenley Park Conservatory.

The group of Begonias on page 120 gives a fair idea of the variety and decorative effect of these plants, while in Plate I. and II. characteristic leaf forms are represented.

BEGONIA LEAVES.

PLATE II. (SEE NEXT PAGE.)

1. Flora Hill.
2. M. Paraert.
3. Louis Closson.
4. Souvenir de Joseph Main.
5. Nickle Plate.
6. Smaragdina, a small but exquisite leaf.
7. Lady Slade.
8. Clementina.
9. Diadema.
10. Conchaefolia, shell-leaved.
11. Feastii.
12. Compta, adorned; a Brazilian species.

BEGONIA LEAVES. Plate No. 2.

CHRYSANTHEMUMS.

The "Golden Flower;" this is the meaning of the botanical name, being derived from the Greek *chrysos*, gold ; and *anthemon*, a flower. Some of the natural species are, indeed, of a golden yellow and all of them, when not changed by cultivation, have yellow disks. The Chrysanthemums belong like the Cinerarias to the Composite family, and what appears as a single flower to the uninitiated is really a collection of hundreds of flowers. Take one of our common Ox-eyed Daisies and examine it with a good magnifying glass—you will find that the many yellow parts which compose the center or disk, and which look so much like stamens, are in reality little flowers, each with its tubular corolla, its five stamens grown into a tube around the pistil, the stigmas of which protrude at the top and what appears to be white petals are also flowers—the ray or strap-shaped flowers. This Daisy is nothing else but a Chrysanthemum, its botanical name being CHRYSANTHEMUM LEUCANTHEMUM, white-flowered.

But the Chrysanthemums which have become famous for their variety and beauty are imported from China and Japan, where they have been cultivated and improved for centuries. English gardeners have learned the trick of producing new varieties many years ago, and also American florists have gained a reputation in originating new and striking forms. Mr. Wm. Hamilton, of Allegheny Conservatory is the originator of several of them ; for the variety named by him MRS. ANDREW CARNEGIE, he received in 1888 the silver cup prize from the New York Horticultural Society, also a silver medal by the Pennsylvania Horticultural Society ; for his seedling variety named MRS. HENRY PHIPPS, he was awarded a silver cup at the Pittsburg Exposition in 1892.

The Chrysanthemums so much admired at the annual shows are nearly all derived from one species,

Chrysan'themum sinen'se, Chinese ; the varieties are innumerable ; they may be arranged into the following groups :

The INCURVED-FLOWERED ; in these the petals are all strap-shaped and curved towards the center.

The QUILLED or ASTER FLOWERED ; in these the disk flowers are tubular, forming a convex, cushion-like head.

CHRYSANTHEMUMS.
Allegheny Conservatory.

POMPONE FLOWERED; small, but many heads; the flowers reflexed, also fringed or toothed at their tips.

LARGE FLOWERED JAPANESE, with long, loose, narrow and twisted petals.

QUILLED JAPANESE; flower heads from 6 to 9 inches in diameter; petals long, rolled up or tubular, with toothed tips.

To write a list of all the varieties would be of little use; from the preparations made in both Conservatories for the Chrysanthemum exhibit, a display of surpassing grandeur may safely be predicted; a fitting conclusion of the floral year.

Chrysan'themum frutes'cens, shrubby, known as Paris Daisies. French Marguerites should be mentioned here; they are popular greenhouse plants, bloom nearly all the year round and are also found in the Conservatories.

VARIOUS OTHER FLOWERS.

Among the various other flowering plants distributed in different departments of the Conservatories are the following:

Muehlenbeck'ia rotundifolia, named after Dr. H. G. Muehlenbeck, a Swiss physician. [A.]

Muehlenbeckia platyclad'a, also called

Coccol'oba platyclada, from *kokkos*, a berry; and *lobos*, a pod, in reference to the fruit; *platiclada*, broad-branched. This peculiar plant is now frequently met with; its branches are flat and green-like leaves, and there are often very few real leaves on the plant, as the stem is able to attend to the functions of the foliage, that is, take in carbonic acid gas from the air and from this and the crude sap absorbed by the roots, elaborate vegetable substance. These plants are from the Solomon islands and belong to the Knotweed family (Polygoniacea). [A. & S.]

Illic'ium anisat'um, the Anis-seed tree; from *illicia*, to entice, on account of the agreeable odor of the plant; *anisatum*, anis-scented. The tree belongs to the Magnolia family and grows native in China and Japan; it is held sacred by the Japanese who form wreaths of it with which to decorate the tombs of their deceased friends. The leaves are said to be poisonous.

Camel'lia thei'fera, Tea plant; named in honor of George Joseph Camellus or Kamel, a Moravian Jesuit and traveller in Asia, who wrote of the plants on the Isle of Luzon; *theifera*, thea-bearing. Our most valuable gift from the Mongol race, furnishing the beverage which exhilirates but does not intoxicate, is not only an exceedingly useful, but also a handsome plant, well deserving cultivation, though it cannot rival the beauty of its sister [s.]

Camellia Japonica, the common Camellia, so delicate, so proud, so frigid. [A.]

Abut'ilon vexillar'ium, Flowering Maple; *abutilon*, is the Arabic name; *vexillarium*, standard. A graceful shrub with drooping branches and bell-shaped flowers which have dark red sepals, pale yellow petals and brown stamens. The leaf slightly resembles a Maple leaf. From Rio Grande.

A. Thomp'soni, has large, striated yellow flowers and mottled leaves. Besides these there are some of the varieties in the Conservatories. The Abutilons belong to the Mallow family.

Lin'um flav'um, Yellow Flax. *Linum*, is the old Greek name from which our word linen is derived. The common flax cultivated for its fibre is Linum usitatissimum (most used); it has blue flowers, whereas the species named above has golden yellow blossoms. It is a native of Europe and belongs to the Linum family (Linaceae).

Impat'iens Sultan'i, Sultan of Zanzibar's Balsam. This is one of the many species of Balsam or Touch-me-not, so called because of a peculiar arrangement of the seed pod which causes it to snap open and coil in a spiral when touched ever so lightly; this is one of the many wonderful arrangements by which plants scatter their seed. Two of the Touch-me-not's grow abundantly in moist shaded ravines around Pittsburg (Impatiens pallida and I. fulva, the pale and the tawny), The above named species is from Zanzibar and has handsome scarlet flowers with slender spurs. They belong to the Geranium family. [s.]

Melia Azadirachta, Bead Tree. *Melia*, is the Greek name for ash; the species name is probably an Indian name. The specimen at Schenley Park Conservatory is at present a small shrub; it is called Bead Tree because its seeds have a natural perforation through the center and are used in Catholic countries for rosary beads. It is also called Arbor Sancta. It comes from the East Indies and belongs to the Melia family which is allied to the Geranium family.

Euphor'bia splen'dens, Crown of Thorns. This genus is said to be named after *Euphorbus*, a physician to Juba, King of Mauritania. There are many species, growing in various climes; many of them have medicinal properties, mainly purgative, hence the common name of Spurge. E. Splendens is one of the most interesting, its stem is full of prickles and often twisted and coiled ; sometimes it is entirely bare of leaves but studded with small blood-red flowers; hence the name " Crown of Thorns " is well chosen. Its home is the island of Bourbon. [A. & S.]

Euphorbia pulcher'rima, pretty, has large vermillion bracts under the rather small yellow flowers. Also named Poinsettia pulcherima. From Mexico. [A.]

Acalyph'a musa'ica, Nettle Spurge. *Acalypha*, is the name given by Hippocrates to the Nettle ; *musaica*, mosaic. A plant belonging, like the above, to the Spurge family (Euphorbiaceae) and resembling a Nettle in appearance. Its bronzy green leaves, variegated with orange and dull red make it an attractive plant. From Polynesia. [A.]

Acalypha marginat'a, has large very hairy leaves, brown in the center and with a rose-colored margin. From the Fiji islands. [A.]

Acalypha Macafeean'a, nas red leaves blotched with bronzy crimson. [S.]

Aral'ia Guilfoyl'ei. This plant of the Ginseng family (Araliaceae) comes from the South Sea Islands ; like many other plants of this genus it is valued by gardeners on account of its graceful pinnately compound leaves. [A.]

Pan'ax Victoria, Queen Victoria's Ginseng. The genus name is derived from *panakes*, a panacea, the ancients believing that it was a remedy for all complaints. It also belongs to the Ginseng family and is largely cultivated on account of its graceful, much divided and prettily variegated leaves. [A. & S.]

Eugen'ia triternat'a, Cambuy Fruit. Named in honor of Prince Eugen of Savoy, who possessed a botanical garden and did much to encourage the study of plants; *triternata*, three times four compound, the leaf having 81 leaflets. These plants belong to the Myrtle family ; the best known of the genus is Eugenia pimenta, which furnishes the Allspice. [A.]

Centraden'iarosea, from *kentros*, a spur ; and *aden*, a gland, referring to the spur-like gland on the anthers. A plant with clusters of pink flowers, of the family Melastomaceae and native of Mexico. [A]

Saxi'fraga sarmentos'a, Old Man's Beard, Aaron's Beard, Creeping Sailor; *saxum*, a rock; *frango*, to break. Many plants of the Saxifrage family grow in rocky places, the roots penetrating and widening the fissures between the stones; hence the name. Some authors believe, however, that the name indicates a supposed medicinal property of some plants of this genus. The many common names of the plant show that it is widely known; there are few plants prettier if growing in rockeries or in hanging baskets. They are natives of Japan. [s.]

Acac'ia pubes'cens, Acacia. The name is believed to be derived from *akazo*, to sharpen, on account of the spines which many species possess; *pubescens*, downy. The Acacias belong to the Pulse family (Leguminose); they are shrubs or trees and some of the species are of great importance. ACACIA ARABICA, A. VERA, A. ADANSONI, A. VEREK and others furnish the Gum Arabic; they are natives of Arabia and northern Africa. ACACIA CATECHU is the principal plant from which the drug Catechu is prepared. Other species furnish other important drugs, gums also tannin. All of them are characterized by their pinnately, compound leaves, with numerous small leaflets. The species principally cultivated are from Australia and New South Wales. The species named above has yellow flowers of delicious fragrance. Two or three tiny trees that were in bloom last March filled the whole show house in Schenley Park Conservatory with their exquisite odor. [s.]

In'ga pulcher'rima, Fairest Inga. This is another plant of the Pulse family, a Mexican shrub with drooping heads of scarlet flowers and fine feathery compound leaves. [A.]

Ardis'ia crenulat'a, *ardis*, a point; this refers to the spear-pointed anthers; *crenulata*, finely scalloped margins. This is one of many evergreen species of Ardisias which are cultivated for their handsome evergreen foliage; it has reddish violet flowers but is still prettier when the bright coral-red berries are ripe. It is a native of Mexico and belongs to the Myrsina family. [A.]

Ol'ea frag'rans, Fragrant Olive. A handsome shrub from Japan, belonging to the Olive family (Oleaceae); though it is more properly named Osmanthus fragrans, from *osme*, perfume; and *anthos*, a flower. The yellow and white flowers are exceedingly fragrant. The true Olive, Olea-Europea, deserves also cultivation in gardens; it has the aspect of a silvery willow and is deliciously fragrant during blooming time.

Duran'ta Bumgart'ii, named in honor of Castor Durantes, a physician and botanist of the sixteenth century. It is an evergreen shrub with blue flowers, belonging to the Verbena family.

Sal'via splen'dens, Showy Sage. *Salvia*, is the old Latin name, from *salveo*, to save or heal. This is a well known cultivated sage with showy scarlet flower clusters. In Brazil a common native plant. [s.]

Francis'cea confert'ii, better known to botanists under the name of

Brunfelsia confertiflora, named after Otto Brunfels, first a Carthusian monk, afterwards, a physician; *confertiflora*, dense flowered. One of the many interesting plants of the Figwort family (Scrophulariacea) with dense clusters of blue flowers; from Brazil. [A]

Brunfelsia eximia, choice; has deep, purple flowers and is also from Brazil. [A.]

Mimulus moschat'us, Musk. From *mimno*, an ape; the the two-lipped flower having been compared with the face of an ape; *moschatus*, musk. A well known and pretty little plant with clammy leaves and yellow flowers strongly musk scented. Made itself easily at home among the rocks in the Palm house. Figwort family. [s.]

Russel'lia jun'cea, named in honor of Alexander Russell, M. D., *juncea*, rush like. Another plant of the Figwort family with rush-like branches; from Mexico. [A.]

Torenia Fournier'i, named after Olef Toren, a Swedish clergyman, who discovered in Asia the first plants of this genus. Many specimens can be seen in Schenley Park Conservatory and their pretty two-lipped flowers with pale blue tube, lower lip and side lobes of dark velvety violet and an orange spot in the middle, should attract the attention of the visitor. From Cochin China. [s.]

Torenia Bailloni, also called T. flava (yellow), has a flower with a brown tube and yellow lips. These, too, belong to the Figwort family and resemble the Monkey flower (Mimulus). They are well adapted for floral baskets. [s.]

Coffe'a arab'ica, Coffee tree. The name is taken from Coffee, a province of Narea, in Africa where the Coffee grows in abundance. The Coffee tree belongs to the Madder family (Rubiaceae) and the specimens in both Conservatories are not only interesting objects to the visitors, among whom there are many who dearly love the infusion prepared from the roasted bean, but they are also attractive on account of their dark, shiny, evergreen leaves and their clusters of white, sweet-scented flowers.

Garden'ia florida, Cape Jessamine, named in honor of Dr. Alexander, of Charlestown, South Carolina, a correspondent of Linnæus. A well known shrub with fragrant white flowers, excellently suited for bouquets ; a native of China, belonging to the Madder family. [A.]

Hoffman'ia Higgin'sia, named after Prof. G. F. Hoffmann in Goettingen. The shrub is also called Campylobotrys Higginsii; belongs, like the above, to the Madder family and is a native of tropical America. [A.]

Hebeclin'ium ian'thenum, or

Eupator'ium ianthenum. According to Pliny the name is derived from Mithridates Eupator, king of Pontus, who discovered one of the species to be an antidote against poison ; *Ianthinum*, violet. A profusely flowering herb of the Composite family, with large clustered violet flowers and soft, large leaves ; frequently cultivated in greenhouses. The Boneset or Thoroughwort and several other medicinal plants belong to the Eupatoriums. [A.]

BEDDING PLANTS.

Some houses in both Conservatories are occupied in winter with plants which are set out during the summer, furnishing the material for the much admired decorative flower beds. The more showy of them, such as the Cannas, Geraniums and others have already been discussed ; the smaller, but not less important plants which are used to frame the larger groups with pretty borders, shall have at least a passing notice ; they are the curly

Alternanther'as of the Amarant family which owe their name to the fact that the anthers are alternately barren. As the flower is quite inconspicuous, this feature would scarcely be discovered by any one but a botanist armed with his Coddington lens.

Who does not know and who does not grow the

Col'eus plants with their square stems and opposite leaves of most varied and richest colors ; the name again has reference to a feature of the flower not easily detected by the uninitiated in floral anatomy; the filaments of the stamens are grown together, forming a tube which surrounds the slender style like a sheath holds the sword; the Greek for sheath is *koleos*, hence the name of the plant. It belongs to the Mint family (Labiatae). The form most largely used for bedding purposes is

Coleus Verschaffelti, a variety of Coleus Blumii, which is a native of Java.

Achyran'thes, a synonym for

Iresin'e, *eiros*, wool, referring to the wooly aspect of some species. These have some of the aspect of Coleus though they belong to the Amaranth family; they are chosen on account of the deep glowing red colors of their leaves.

I. Linden'ii, with deep vineous red stems and leafstalks and

I. Herbstii, the leaves of which are two-lobed of the apex, upper surface dark maroon, midrib and principal veins carmine; these are the species most largely used.

Centaur'ea, a very different plant from the one just mentioned, but its silvery or dusty whitish leaves form a beautiful contrast to the glowing colors of the Coleus. The name is derived from *Kentaurion*, the name of the plant which according to Ovid's fable cured Chiron, one of the Centauri. The Centaureas belong to the composite family.

C. cineraria, ashen, and

C. ragusina, from Ragusa, in Dalmatia, are the species mostly cultivated for bedding purposes, while

C. cyanus, blue, the Corn Flower, so pretty and so abundant in the grainfields of Europe, belongs in the button hole of an unmarried man who ought to be married.

Santolina, Lavender Cotton, is another composite plant from the Mediterranean region; its name is said to be derived from *Santonia*, a kind of wormwood, according to Pliny. The pretty little shrub used in the borders of some beds in front of Schenley Park Conservatory is no doubt SENTOLINA CHAMAERYPARISSUS INCANA; but SANTOLINA RASMARINIFOLIA is also very well adapted for decorative bedding.

Lobel'ia pum'ila magnif'ica. Magnificent Dwarf Lobelia, is one of the several varieties much used as carpet and border plant; the name Lobelia commemorates Mathias de L'Obel, a botanist and physician to James I.

Nierembergia gracolis, named after John Nieremberg, a Jesuit who wrote a book on the marvels of nature. This is a graceful plant with pale blue bell-shaped flowers. At this writing there is a bed of these pretty flowers right near the entrance to the park near the bridge.

Mesembryan'themum, Fig Marigold ; from *mesembrya*, midday ; and *anthemum*, a flower, so named because the flowers open only in bright sunlight. Who does not know these pretty little almost moss like plants with small but fleshy leaves? There is scarcely a flower box or hanging basket without them. They are mostly South African plants of the Fig Marigold family (Ficoidaceae).

Echever'ia, also called COTYLEDON ; the first name was given in honor of M. Echeveria, botanical draughtsman ; the other is derived from *kotyle*, a cavity, alluding to the hollow leaves of some species. They belong to the Houseleek family (Crassulaceae) and are characterized by their very thick fleshy leaves which grow in the form of rosettes. For horticultural sculpturing there is no more serviceable plant, the dense and rather slow equal growth makes it possible to execute designs which keep unchanged for a considerable time, while the different kinds of green of the leaves of different species or varieties, make it possible to the skillful gardener to give his picture the lights and shades which help so much to impart expression and animation. Visitors of the Allegheny Parks have admired in former years the medallion reliefs which Mr. Hamilton had executed and which were made entirely from Echeverias. This year again the head of Lincoln and the emblems of the Grand Army of the Republic can be seen. The sight will no doubt gladden the hearts of the visiting veterans. A representation of Lincoln's head can be found on page 137. It is entirely made of Echeveria glauca and secunda, while Alternantheras furnish the colors for the other designs. There are also fine specimens of plant sculpturing in Highland Park.

Corps Badge
Army of Tennessee.

Lincoln.

Corps Badge
Army of the Potomac.

Memoranda.

CLIMBING PLANTS.

Kneeling Venus Framed in
Solanum Seaforthianum.
Schenley Park Conservatory.

Climbing Plants.

The honest forest trees build up their structures carefully; every year the roots are anchored more securely; a new layer is added to the trunk, so that the expanding crown of branches may have a firm support, a secure foundation.

There comes a nimble climbing plant; it installs itself close to the stately trunk; it sinks its grasping rootlets or tendrils into the cracks of the bark; in one short season it has run up the height of the stem, in a few more it has reached the top of the crown and spread over all the branches, basking in the sunlight, drinking up the rain, enjoying all the advantages for which the poor tree has toiled a century and of which he is now, to a great extent, deprived. Often the honest host is sapped, strangled, suffocated, by his unbidden guest, who adorns now the ruin he has wrought, until the murdered giant crumbles into dust and his assassin creeps on along the ground in search of a new victim. This is one of the many scenes of strife in the forest.

Not all the climbing and twining plants elevate themselves at the expense of some unfortunate victim; some make use of their peculiar powers, to climb cliffs and crags or to descend from rocky precipices, clothing with ornamental verdure the most inaccessible places.

Man who presses everything into his service has also turned the faculties of those plants to good account; how they are made to clothe the walls of our houses, to shade our verandas, and adorn our corridors, the reader well knows.

In many respects climbing plants are an interesting study; one of the most fascinating chapters of that study is the investigation of the various contrivances by means of which the plants can raise, attach and support themselves. To those interested in such researches the author would suggest to notice the manner in which the English Ivy fastens itself by means of little rootlets, the Virginia Creeper, the Boston Ivy (Ampelopsis Veitchii), the Grapevine, each of which supports itself by leaves, modified in a peculiar manner;

the Clematis, Honeysuckle, Moon vine, even the Sweet Pea and Morning Glory are interesting objects of that study. In the Conservatories there ia a chance to observe the manner of climbing of many tropical plants. The passion vine should not be overlooked. There are some families of plants in which the habit of climbing is characteristic of all or most of the species, as in the Bindweed family (Convolvulaceae). The Grapevine family (Vitaceae), the Birthroot family (Aristolochiaceae) and others. But there are many other families in which only one or a few species have taken to climbing, as the Climbing Rose in the Rose family, the Lapageria in the Lily family, the Mikania in the Composite family and others which will be mentioned. Even among the Orchids, the Palms and the Ferns we find climbers.

The following is a list of climbing and twining plants which may be seen at present in the Conservatories.

Glorios'a Superb'a. The name suggests a glorious flower which this interesting Lily indeed produces. A most peculiar feature in this plant are the tendril-like tips of the leaves by means of which the plant can climb and find support among the surrounding vegetation. It is a native of tropical Asia and Africa. [s.]

Lapager'ia ros'ea, named after Josephine Tascher de la Pagerie, who became the wife of Napoleon I. and who was an ardent botanist. Few climbing plants are more graceful than the Lapagerias. They are especially adapted for training over corridors, as here the pendant, waxy flowers are seen to the best advantage. The species L. rosea has rich, rosy crimson flowers, produced in great abundance and remaining in full beauty for several months. They are natives of Chili. [s.]

Lapageria rosea alba, has pure white flowers. [s.]

Antig'onous lep'topus, *anti*, opposite; *gonia*, an angle; *leptopus*, slender stemmed. A fine climber from Mexico; especially attractive in flowering time when it bears a profusion of flowers, each shaded from light to deep pink. Knotweed family. (Polygoniaceae).

Bougainvil'læa glab'ra. Named after the French navigator De Bougainville; *glabra*, smooth. Few climbing plants are more gorgeous than these inhabitants of the Brazilian forests. The flowers are not conspicuous, but the bracts surrounding them are beautifully colored, being pale rose-tinted in the species named above while in

B. specios'a, beautiful, they merge into rose-lilac.

B. spectab'ilis, showy, is seldom seen in bloom.

B. refulgens, shiny, has brilliant purple-mauve bracts, produced in long, pendulous racemes. These plants belong to the Four o'clock family (Nictaginaceae).

Ledenbergia rosea=æna, Climbing Poke. Named after Mr. Ledenberg; *rosea-æna,* rosy bronzed. This is a South American plant of the Poke family (Phytolacaceae). The flowers are inconspicuous, but the foliage is handsome, of a dark shining coppery green on the upper surface and bright, rosy violet color beneath.

[s.]

Aristoloch'ia grandiflora, Large-flowered Dutchman's Pipe. *Aristos,* best; *locheia,* parturition, referring to supposed medicinal properties, hence also the common name Birthroot. The Dutchman's Pipe needs no introduction; its peculiar flower is always an object of curiosity; as a climbing plant, shading a veranda or a cozy bower, few vines are better adapted; the large heart shaped leaves arrange themselves like shingles and very effectually protect those behind them from the gaze of curious neighbors or passers-by. The species named above is not the only one with large flowers; there is also

A. gig'as, giant, with large purple flowers, the tube inflated and contracted in the middle and with a wide border at the top; this species is from Guatamala. [s.]

A. elegans, is from Brazil and has very odd flowers.

A. ridicu'la, ridiculous, also from Brazil, has immense flowers from 3½ to 4½ inches in length; the tube is almost doubled upon itself and the border has two long expansions named by somebody "Donkey's Ears." The coloring of the flowers and the tracery of the veins are also very peculiar. These are tropical vines and cannot be raised outside of the Conservatory. The most common and probably best hardy climber among the "Pipe Vines" is Aristolochia Sypho which is found in the Allegheny mountains. Birthroot family (Aristolochiaceae).

Pip'er rotundifol'ia, Pepper vine. *Piper,* is the old name for pepper; *rotundifolia,* round-leaved. The black pepper in our spice boxes is the fruit of Piper nigrum; white pepper is the same with the outer shell removed. [s.]

Paullin'ia thalictrifolia. Named after Ch. Fr. Paulli, a Dutch Botanist; *thalictrifolia,* Meadow-Rue-leaved. A beautiful evergreen climber with many times compound leaves and pale pink flowers; from Rio de Janeiro.

Cis'sus dis'color, Painted Vine, *kissos,* Ivy, so called on account of its climbing habit; *discolor,* variously colored. This is certainly one of the most beautiful climbing plants; its variegated leaves showing exceedingly fine color combinations; it is from Java and belongs to the Grapevine family (Vitaceae). The visitor may see this plant, on passing from the Palm house in Schenley Park Conservatory to the tropical house at the right, where there is, at least at present, a fine specimen at each side of the door. [s.]

Passiflora, Passion flower. Missionaries who found this vine in South America gave it the now well known name, because they imagined to discover in the blossoms a representation of all the implements of crucifixion. The flower is certainly a strange and beautiful object. The passion vines are mostly American, some species being quite common in the South; a few are found in Asia and Australia; the Conservatories contain some of the finest specimens, as

Passiflor'a arbor'ea, tree-like, from New Grenada.

P. cærul'ea, blue-flowered; the fragrant flowers last but one day. From it is derived the fine variety

P. cærulea Constance Elliott, which has white flowers. [s.]

P. prin'ceps, princely, with deep red or scarlet flowers; from Brazil. [A. & s.]

P. Pfor'tii. [A. & s.]

Tacson'ia insig'nis, *Tacso,* the Peruvian name of one of the species; *insignis,* remarkable. This plant also belongs to the Passiflorae and resembles the Passion vine. From Peru. [s.]

T. exoniensis. [s.]

T. Van Volexem'ii, this is one of the finest species and has very showy scarlet flowers. From New Grenada. [s.]

Swainson'ia galegifolia. Named in honor of Isaac Swainson a celebrated cultivator of plants. These plants of the Pulse family (Leguminosae) are exceedingly graceful, on account of their slender stems, finely divided leaves and delicate flowers. They are natives of Australia and will find a place soon among the most popular plants in the florist's show window. The above species has deep red flowers, while

S. Osborne, has pure white blossoms and there are other species and varieties representing all, or nearly all, the delicate tints which we find in the Sweet Pea, which the flower much resembles.

[s.]

Solanum, this is the old Latin name for the Nightshade; the Nightshade family (Solanaceae) contains many very important plants; some useful, as the Potatoe, Tomato, Egg plant; some noxious as the Tobacco, Belladonna, Henbane, Stramonium and other poisonous plants. Many of the Solanums are climbers and the handsomest of these is certainly

Solenum Seaforthian'um, which is represented in the frontispiece to this chapter. This graceful climber, which clothes the wall of the office and forms an elegant frame for the "Kneeling Venus" is admired by every visitor; it seems never to tire in sending out its cluster of handsome flowers. It is a native of the West Indies. [s.]

S. jasminoid'es, Jasmine-like, has white flowers and comes from South America; it is frequently met with in the greenhouses.

[s.]

S. Guatemaliensis, is a new species received by Mr. Hamilton for the Allegheny Conservatory; also

S. Species unknown, from New Holland; the plant was exhibited in Chicago as "Kangaroo Apple," it has white flowers much like those of the potatoe, and bears a rather large berry which is said to form a favorite food of the Kangaroos. Another plant which has the same common name is S. AVICULARE, which has blue flowers. [A.]

Habrotham'nus el'egans, a synonym of

Ces'trum elegans, *kestron*, the Greek name for another plant. This plant with purplish red flowers also belongs to the Nightshade family; it comes from Mexico. [A.]

Ipom'ea Horsfal'liæ, Mrs. Horsefall's Bindweed. *Ips*, Bindweed; *omoios*, similar; so called on account of the resemblance of these plants to the Bindweed or Morning Glory. The Bindweed family (Convolvulaceae) is one in which the habit of climbing and twining is a common characteristic; to it belong our wild and cultivated Morning Glories, the Moon-vine, the Sweet Potatoe and other well known plants. The above named species is a fine evergreen twiner with hand-shaped leaves and rich, deep pink flowers; from the West Indies. [s.]

Cobæa scan'dens. Named after the Spanish botanist B. Cobo; *scandens*, climbing. A very well known climbing plant, cultivated everywhere; its popularity is due to its rapid growth and its beautiful bell-shaped flowers. It is a native of Mexico and belongs to the Phlox or Polemonium family. [s.]

Hoy'a carnos'a, Wax Flower. Named after Thomas Hoy, once gardener to the Duke of Northumberland; *carnosa*, fleshy. This is another plant which needs no introduction to the reader. The Wax flower, with its shiny, evergreen leaves and its umbels of delicate flesh-colored flowers, which look indeed as if they were formed of wax, is grown in many a home. It belongs to the Milkweed family (Asclepiadaceae) and comes from Queensland.

Physian'thus al'bens, White Bladder Flower. *Physa*, a bladder; *anthos*, a flower; *albens*, white. This is a Brazilian climber belonging, like the above, to the Milkweed family.

Stephanot'is floribun'da, Clustered Wax Flower, Madagascar Jasmine. *Stephanos*, a crown; *out, otos*, an ear, alluding to the auricles on the crown of the stamens; *floribunda*, bundle flowered. This is certainly one of the most elegant of climbing plants; at flowering time its large clusters of pure white flowers fill the Conservatory with sweet fragrance. Belongs to the Milkweed family and comes from Madagascar. There is no objection to using a sprig of this lovely plant in the bridal wreath.

Jasmin'um grandiflorum, Large flowered Jasmine. One of the many cultivated Jasmines, a favorite outdoor plant in the tropical and sub-tropical countries; it has white flowers, reddish underneath and grows wild in the Himalaya mountains. Belongs to the the Olive family. Although not exactly a climbing plant it is mentioned here because so many of the Jasmines are climbers. [s.]

Jasmin'um Sambac, Zamback or Arabian Jasmine, has fragrant white flowers; it is a handsome twiner from the East Indies and blooms nearly all the year round. [A.]

Allaman'da grandiflor'a. Mr. Allamand of Leyden sent some seeds of a plant of this genus which he found in northern South America to Linnaeus who returned the compliment by naming the plant after him. This and

A. Henderson'y, a variety of A. Cathartica, are beautiful climbing plants of the Dogbane family (Apocinacea). From South America. [s.]

Rhyncosperm′um, a synonym of

Trachelosperm′um jasmonides, *Trachelos,* the neck; and *sperma,* a seed, referring to the prolongation of the termination of the seed; jasminoides, Jasmine-like. This is a white flowered and very fragrant climber from Shanghai. [s.]

Cleroden′dron Balfourianum, *kleros,* chance; *dendron,* a tree; so called, it is said, from the uncertainty of its medicinal qualities. No one who ever visited Allegheny Conservatory can have overlooked this splendid climbing plant which covers the whole wall at the entrance to the Palm house. There are also some fine plants in Schenley Park Conservatory. When in full profusion of its scarlet flowers with white calices it forms a gorgeous adornment of a wall. They are natives of Asia and the above is a variety of

Clerodendron Thomson′ae, named after a Mrs. Thomson. The reader will be surprised to learn that this plant belongs to the Verbena family (Verbenaceae). [A. & s.]

Hexacen′tris mysorensis, a synonym for

Thun′bergia mysoren′sis (See page 70). A rather rapidly growing and spreading climbing plant with large purplish and yellow flowers; from Mysore. Acanthus family. [s.]

Bignonia venusta. Named after Abbe Bignon, librarian to Louis IV; *venusta,* handsome. The Bignonias are grand South American climbing plants with immense yellow, orange, red, purple, also white flowers. In their native tropical forest where they can roam from tree to tree expanding their rich foliage and their gorgeous blossoms, they must present a glorious sight. B. venusta has funnel-shaped crimson flowers. [s.]

Another splendid climber of the Bignonia family is

Tec′oma rad′icans, Trumpet Creeper. From the Mexican name, *Tecomaxochitl; radicans,* rooting. Everybody knows this thrifty vine with bright red flowers; it grows wild in the Alleghenies in the southern part of Pennsylvania and all through the mountainous districts of the South. [s.]

T. grandiflora, large flowered, from Japan and China and

T. jasminoides, Jasmine-like, from Australia, can also be found in Schenley Park Conservatory and should be inspected by the visitor.

Manet'tia bi'color, named after Xavier Manetti, Prefect of the Botanic Gardens at Florence, in the middle of the last century. This rather delicate climber has tubular flowers, bright scarlet below, golden yellow near the opening; it belongs to the Madder family and has its home in the Organ mountains. [A. & S.]

Mican'ia violac'ea, named after Prof. Joseph Mican, of Prague. One of the few climbing plants found in the Composite family; it bears clusters of handsome violet flowers and hails from South America. [S.]

AQUATIC PLANTS.

ALLIGATOR POND. Schenley Park Palm House.

AQUATIC PLANTS.

Water has its vegetation as well as the land. The bottom of the sea in some parts is said to resemble a flower garden; not only rendered so by the profusion of delicate and coarse seaweeds of rich and varied colors, but also on account of the flower-like animals, the sea anemones, the corals, moss animals, sea ferns and other phantastic creatures. Every pond, lake and river has its vegetable life and especially in the tropics this life is abundant and multiform. The aquatic departments in the Phipps Conservatories are well planned and well equipped and give facilites for the rearing of the rarest of aquatic plants.

These ponds and reservoirs add quite a charm to the greenhouses; a quiet peace seems to rest upon the waters which even the little goldfish, playing between the floating foliage, do not disturb. An imaginative mind may dream strange dreams in these surroundings.

The genius of fancy may stealthily take possession of you; the pond grows larger and larger, until before you lies a placid lake; its shores, with many a bend and nook, borders upon a scene of topic aspect; this side is rendered dark by mighty trees hung with giant climbing plants, whose festoons dip down into the wave; yonder shore is lined by Papyrus, gracefully bending to see their plumaged heads reflected in the water; again you spy large groups of Lotus flowers lifting the velvety shields well above the water, but raising higher still the globe-shaped flowers of delicate tint and perfume.

Like floating islands appear the patches of Water Lilies, their circled leaves spread out upon the glassy surface; their flowers so exceeding fair, of purest white or purple, rose or scarlet, dreamily looking at the starry sky, while the still air is saturated with delicious fragrance. The moon, peeping down between the tree tops, throws a trembling, glittering bridge across the water; and there

> Glides a canoe
> With room for two—
> 'Tis he and you.

But now I must leave it to the fair reader to continue and complete the vision.

VICTORIA REGIA.
Allegheny Conservatory.

It is not long since it has been discovered that the Lotus flower and the pink and blue Water Lilies could be cultivated out doors in this country and since that time many a swamp and ugly pond in New Jersey and down through Maryland has been changed into an almost paradisical spot. But some of the tropical water plants in the Conservatories can only be raised when the utmost care is taken that a certain range of temperature of the water and the air is maintained.

WATER LILIES.

These lovely plants, which are by no means strangers to this clime, as those who have visited any of the lakes of this State, but especially the lakes of northern New York and New England can attest, form a family of their own; they are botanically quite separated from the other Lilies which have been discussed in the chapter on flowers. One of the most interesting features to the student of forms of plant organs, a study called Morphology, is the gradual merging of the petals into stamens. The acknowledged queen among the Water Lilies is the

VICTORIA REGIA.

A special large reservoir is reserved for this plant in each of the Conservatories. The temperature of water and air is most carefully regulated; starting from the seed in spring, it can be held in a saucer; two months later its leaves cover the whole reservoir, some of the circular blades measuring more than 6 feet in diameter. Leaf after leaf is sent out in regular two-fifth ranking arrangement until the large reservoir becomes too small and the older leaves have to be constantly removed. During the most rapid stage of growth leaves add in 24 hours from 20 inches to 2 feet to their diameter.

Nothing is more wonderful than these gigantic circular leaves with upturned edges. Its most interesting part is its venation; this can only be seen when the leaf is turned upside down. From the center, where the petiole is fastened, strong cylindrical ribs, thickly beset with prickles, radiate in every direction, each rib dividing into two smaller ribs, and each of these subdividing in the same manner. Between these ribs are numerous cross partitions, forming compartments in which air seems to be held, for indeed the leaf layer proper is supported above the surface of the water. The petioles and the largest ribs are provided with two large air-ducts, fully one-eighth of an inch in diameter, and several smaller ones.

Leaf of Victoria Regia, showing the venation on the under side; to the right a group of Papyrus antiquorum; at the left, in the background, leaves of Nelumbium speciosum (Lotus flower), Allegheny Conservatory.

When once beginning to bloom it generally continues blooming for considerable time. The flowers measure about 12 inches in diameter; they are pinkish when they open; they become pure white and again change into a dark pink as they fade. The fruit is ripened under water; it consists of a large globular berry, beset with prickles.

The seed is rich in starch and forms an article of food in the home of the plant, Brazil. (Hence also the name "Water Maize").

Victoria Randi, the new Crimson Victoria, is a variety of recent introduction; very similer to the Victoria Regia, except the vertical edges of the leaves are broader, forming a deeper 'tray' and the flowers, opening white, soon change to a deep crimson.

Eury'ale fer'ox, Indian Water Lily. *Euryale* was one of the Gorgons, represented with fierce thorny locks; *ferox*, fierce. This gorgeous plant was considered the noblest of the Water Lilies before the Victoria Regia was discovered; its leaves resemble those of the the Victoria, but are not as large and do not form an upright rim. The under side of the leaves is dark purple and the veins as well as the leafstalk and the calyx are beset with sharp spines; this suggested the scientific name of the plant. It is a native of the East Indies.

[S.]

Nymphaea, Water Lily. The name is derived from nymphe, a water-nymph; horticulturists sometimes distinguish between night-blooming and day-blooming Water Lilies, this distinction is here observed.

The poet Moore in "Paradise and the Peri" hints at the habit of the day-blooming Water Lily, in the following lines :—

> Those Virgin lilies, all the night
> Bathing their beauties in the lake
> That they may rise more fresh and bright
> When their beloved sun's awake.

Among the night-blooming species the following may be found in the Conservatories:

Nymphæ'a lot'us, Egyptian Lotus, the white Lotus of the Nile, has large white flowers with red-margined sepals and sharply toothed leaves; the veins are quite prominent on the under side. This is the sacred Lotus of the ancient Egyptians, dedicated to Isis and is found engraved on very ancient coins. The plant grows abundantly in the plains of lower Egypt during the time the land is under water. Like the Victoria it is useful as well as beautiful; at least the old Egyptians made use of it, drying and grinding the seeds and making bread of it; also the roots which contain a great deal of starch were utilized as food.

N. dentat'a, toothed; is a variety of the above with very large blossoms, from 6 to 19 inches in diameter and with large toothed leaves; from Sierra Leone.

N. Devonien'sis, named after the Duke of Devonshire, has brilliant rose colored petals; it is a hybrid between N. lotus and N. rubra; the flowers are raised a foot and more above the water.

N. rubra, red, resembles the one just mentioned; it is from the East Indies.

The following are novelties from the establishment of Mr. Wm. Tricker of Clifton, N. J., who makes a specialty of raising aquatic plants; he received several prizes at the Columbian Exposition:—

Nymphæ'a Columbian'a, deeper in the color of flower and leaf than N. Devoniensis.

N. Deanian'a; a robust plant with large, deep-green, dentated foliage; the under surface peculiarly blotched; the flower is cup shaped, sepals deep rose pink, stamens red.

N. delicatis'sina, most delicate; has large handsome foliage of metallic luster and large flowers of a delicate pink.

N. Tricker'ii, one of the most magnificent novelties; the flowers are rose pink, suffused with white, becoming darker the second day; the leaves have a metallic reddish-green color.

N. Sturtevan'tii, resembles the above in shape and color of flower, it is also a garden hybrid of American origin, with large bright red flowers and brownish leaves of a metallic luster. It is derived from N. Devoniensis.

Among the day flowering Water Lilies we find the following:

Nymphæa Zanzibaren'sis, Zanzibar Water Lily. One of the most beautiful of all Water Lilies and exceedingly fragrant; the flower is intense blue; the anthers with a shade of violet; sepals green outside, purple within. Several varieties are derived from this.

N. Zanzibarensis azur'ea, azure blue; several shades brighter than the above.

N. Zanzibarensis ros'ea, has a deep rose flower.

N. cærul'ea, also called N. stellata, has light blue petals very delicately scented.

N. scutifol'ia, shield-leaved; another blue flowering species; the leaves are purplish tinted and spotted underneath. From the Cape of Good Hope.

N. Mexican'a, has rich golden flowers; native of New Mexico.

N. el'egans, is also from New Mexico. The flower is white, tinged with purplish blue, fragrant; stamens yellow, tipped with blue.

N. grac'ilis, a graceful, slender plant with greenish white flowers; also from Mexico.

N. gigan'tea, has large blue flowers, seven inches across, with numerous petals and a dense mass of golden yellow stamens. From Australia.

The following are hardy species which stand our climate well.

Nymphæa odorat'a, this is our American fragrant Water Lily especially abundant in lakes, ponds and marshes near the Atlantic coast.

N. odorata Carolinan'a, one of the handsomest varieties of the above, having rose colored flowers.

N. odorata gigan'tea, is a large flowered Southern variety.

N. odorata rosea, Cape Cod Water Lily; this is the large pink Water Lily sold in the streets of New York, Philadelphia and other eastern places.

N. odorata sulphure'a, has a yellow blossom as the name indicates; though it is a shade lighter than sulphur.

N. alba; English white Water Lily, has pure white flowers.

N. alba candidis'sima; a large flowered beautiful variety of the above.

N. Marliac'ea chromatella; this has clear yellow fragrant flowers with orange stamens, the young leaves are mottled with brown.

N. Marliacea al'bida, white.

N. Marliacea car'nea, fresh-tinted.

N. Marliacea ros'ea, pink; are interesting French hybrids.

N. pygmae'a, dwarf: has white fragrant flowers and is a native of Central and Northern Asia.

N. Laydek'eri rosea, is a French form of the pygmaea type, interesting on account of the change of color the flower undergoes from opening to closing.

Nelum'bium specios'um; Sacred Lotus of India; Nelumbium is the Cingalese name, *speciosum*, showy. This most handsome plant is now so largely cultivated in this country, that it is almost as well known here as in its home.

The large circular, velvety leaves, poised upon stout petioles which reach 4 to 6 feet out of the water, present an elegant picture, and when the large, pink and cream colored, fragrant flowers appear above the mass of shields, the aspect is certainly handsome. We do not wonder that this plant has been an object of admiration and worship among the Egyptians, the people of India, China and Japan.

Strange to say, this plant is no longer found growing wild in Egypt. That it must have grown there in ancient times is evident from the many representations of the flower and the leaf on Egyptian monuments and from the description of Herodotus, who compared the receptacle of the flower to a wasp's nest. Strabo and Theophrastus also mention the plant as a native of Egypt. In India, the spiral fibres with which the leaf-stalks abound are carefully extracted and used as wicks for the ever burning lamps in the temples. The fact that the leaves, plunged into water, will emerge dry, owing to the velvety down with which they are covered, gave rise to the Hindoo proverb, that the good and virtuous are like the Lotus leaf; they may be surrounded by the waters of temptation, but they remain undefiled.

Nelumbium album grandiflorum, is a variety with large white flowers.

N. roseum, is a deep pink colored variety.

N. Kermesin'um, another variety of lighter color.

All of the Water Lilies here recorded are grown in Schenley Park Conservatory and many of them also in Allegheny Conservatory.

Cabom'ba aquatica. Cabomba is the name given to the plant in Guiana. A small but interesting water plant with yellow, not very conspicuous flowers. The submerged leaves are opposite and hand shaped; the floating leaves are alternate, shield-shaped and entire. [s.]

OTHER AQUATIC PLANTS.

Eichhorn'ea Pontederia cras'sipes, Water Hyacinth.
A most peculiar plant, constructed to float upon the water. The leaves have swollen, bulb-like stems which contain air and thus buoy up the whole plant. The cluster of handsome, light blue flowers has some resemblance with the inflorescence of the hyacinth, hence the popular name. The plant belongs to the Pickerel-weed family. (Pontederiaceae). [A.]

E. azur'ea is a Brazilian species with lavender blue flowers, purple in the center and a bright yellow spot upon the blue. [S.]

Limnoch'aris Humholdt'ii, Water Poppy; from *limne*, a marsh and *charis*, beauty.

The lovely yellow flower with dark stamens resembles that of the poppy. It belongs to the Water Plantain family and grows wild in Brazil. (Alismaceae). [A. & S.]

Alis'ma plantag'o, Water Plantain. From *alis*, the Celtic word for water. Grows wild on the boarders of marshes and ponds and is attractive enough to deserve a place in the Lily pond in your garden. [A.]

Sagittari'a montevidien'sis. Giant Arrowhead; from *sagitta*, an arrow, so named on account of the shape of the leaf. Some of the arrowheads grow also wild in this country. The above is a large species with white petals, each of which has a crimson spot at the base. Belongs to Water Plantain family and is from South America. [A.]

Aponogeit'on distach'yon. Cape Pond Weed. *Apon* like *Alis* is from Celtic and according to several authorities means water; *geitons*, neighbor; distachion, two-spiked. The flowers of this water plant are very odd, being arranged in two spreading spikes bearing pure white bracts, while the stamens are dark-brown. Belongs to the Family Naiadaceae and is from the Cape of Good Hope. [A.]

Myriophyl'lum proserpinacoid'es, Parrots Feather; Water Millfoil. From *myrius*, myriad; and *phyllon*, a leaf. This delicate feathery, water plant spreads so rapidly that, if not checked in its career it soon fills up the whole basin in which it has been introduced; the visitor can easily recognize it by the yellowish-green color of its finely divided whorled leaves. [A. & S.]

Salvinia natans, Floating Salvinia. Named in honor of Antonio Maria Salvino; Professor at Florence in the Seventeenth Century. Resembling the Duck-weed of our ponds; very small plants with round leaves, floating upon the surface of the water and spreading rapidly; the little leaves are beset with peculiarly formed hairs which hold between them bubbles of air when the plant is pushed below the waters surface, which causes it immediately to rise again. [A.]

All these plants as well as the hardy Water Lilies can be easily reared and a picturesque feature could be added to many a garden by establishing a pond or tank for the raising of some of these beautiful aquatics.

RUSHES.

Papyr'us (Cyper'us) **antiquor'um,** the Ancient Paper Plant This is a tall Sedge with a culm growing 8 or 10 feet in height and one inch in thickness, crowned with an umbel of flowering spickelets, subtended by a close spiral of long slender bracts. The Papyrus grows wild on the banks of the Nile and the Jordan, also in Italy; the paper made from it has not been surpassed in excellence and durability by any paper manufactured in modern times The documents written on it have preserved for our time many most important facts of ancient history. This paper has been made by cutting the stems to the required length, splitting them open lengthwise and peeling off thin slices between the center and the outside. These long slices were laid side by side, and after being watered and beaten with a wooden instrument, were pressed and dried in the sun. The Egyptians well recognized also the graceful beauty of the plant and used it profusely for the decoration of their temples. [A. & S.]

Cyperus alternifol'ius, Alternate-leaved Rush. This Australian species resembles the former, but is smaller. There is also a variegated kind which has the stems and leaves striped with white.

[A. & S.]

Cyper'us Natalensis, a species from Natal. [A.]

Isolep'is grac'ilis, the garden name for

Scir'pus ripar'ius, Club Rush. *Scirpus* is the old Latin name for a Rush; *riparius,* riverside-loving. This rush is very popular for Conservatory decoration and is met with in every greenhouse.

[A. & S.]

THE CACTI.

GROUP OF CACTI. Schenley Park Conservatory.

CACTI.

To seek protection by the trick of offering to the world a bristly exterior is not rare in nature. We find this contrivance in the hedgehog and porcupine, in some fishes, in the sea urchins, in the Chestnut burs and other fruits, in the Rose branch, the Thistle, the Nettle, but most prominently we find it developed among the plants of the Cactus family. It might be said that the human family, too, has its bristly specimens; individuals whom you cannot approach without getting stung or scratched. But behind the dangerous, threatening, spiteful exterior there is often a very harmless creature, satisfied to be let alone and lacking the wit for using his weapons in an aggressive warfare.

The Cacti may not be beautiful, but they are certainly odd and therefore they have their admirers. The raising of these plants has become quite a specialty in floriculture and there are florists who devote themselves exclusively to the raising of these plants. Many Cacti reward the care bestowed upon them by the production of brilliant flowers, put forth, sometimes, in great profusion; in some the fruit is much more conspicuous and showy than the flower; but even when neither flowers nor fruit are in season, these plants present points of great interest to the student of nature. These points are exceedingly sharp, sometimes quite small, sometimes long and stout. Look how they are placed upon the plant, generally in groups on the summit of tubercles, produced in somewhat regular arrangement upon the leaves, or along the edge of parallel, longitudinal or spiral ridges upon the stem. In each group there are thorns of different lengths and thickness and pointing in different directions; but there is a cunning and perfectly symmetrical design in this arrangement and it is repeated in each group of the same plant; and if those groups are placed closely together, the thorns cross each other, forming a perfectly bewildering maze of bristling points, coming from and pointing to every direction, so that it is impossible to take hold of the creature in any manner.

The author had an object lesson on Cacti once. Roaming over an island near the coast of Virginia, he found the first Cactus p'ants (Opuntia vulgaris) growing in their native state. Such a find is an occasion for rejoicing for the botanist. The plants looked rather harmless and a few of them were carefully dug from their sandy bed and placed in the plant press, making rather cumbersome specimens. A few pricks caused by the larger thorns were not minded. But the following morning there was a tickling and burning sensation all over the hands and wrists, which was aggravated by rubbing and which lasted nearly for a week. Many small, almost invisible barbed prickles had penetrated the skin and caused an acute sensation in the cutis.

Two weeks later, having returned home, the plants were examined and seemed not in the least changed, and for fully two months they kept fresh and growing, new joints forming at the expense of the old ones which became soft and brownish and began to rot.

This experience varified the fact often stated, that Cactus plants have an exceedinly thick and dense epidermis which almost prevents the loss of water by evaporation, so that these plants can live through long periods of drought, affording refreshing food to the animals which are plucky enough to tap these springs of the desert with their hoofs.

The Cacti are all natives of the New World, occurring in large numbers and many forms in the northern parts of South America, all through Central America and Mexico, also in the Rocky Mountains and in California.

Who looks at the groups of Cacti in front of Schenley Park Conservatory (during the warm season), need not be told that these plants assume many groteque forms; botanically they are divided into several genera according to certain characteristics of the flower which are in most cases connected with striking differences in the whole aspect of the plant.

The group of Cactus plants on page 162 has been kindly arranged by Mr. Edmonds, foreman of Schenley Park Conservatory, while for the naming of the specimens the auther is indebted to Mr. A. Blanc of A. Blanc & Co., in Philadelphia, who is an expert on Cacti and whose interesting little book "Hints on Cacti" every Cactus fancier can obtain for 10 cents.

In mentioning the different species of Cacti, arranged in groups, which are found in Schenley Park, many of them also in Allegheny Conservatory, those which are represented in the illustration are numbered with the corresponding number.

Cereus, Torch Thistle. From *Cereus,* pliant, in reference to the shoots of some species. From the picture can be seen that many tall, columnar forms are found in this group. They have woody stems with pithy inside, grooved and angled longitudinally, bearing clusters of spines at regular intervals upon the angles and producing large flowers, some of them of exceeding beauty. Several of the species are known under the name of Night-Blooming Cereus, but the most gorgeous of them is Cereus grandiflora, mentioned below. The flowers of the Cereus have a long tubular calyx, the lowest sepals being dark colored, awl-shaped to linear bracts, while farther up they change by degrees into flat, petal-like, light colored divisions, of which there are a great many; the stamens are numerous with long, slender, curved filaments and the long style in the center ends in a star-shaped stigma. Do not fail to examine one of these wonderful flowers if ever you have a chance to get one. The home of the Cereus is South and Central America.

Cer′eus gemmat′us, budding. (1)

C. variab′ilis. (2)

C. Jamacar′ii. (3)

C. Peruvian′us. (4)

C. grandiflor′us, Queen of the Night. (5)

The flowers begin to open between seven and eight o'clock in the evening and are fully expanded by eleven; by three or four o'clock in the morning they fade; but during their short existance there is hardly any flower of greater beauty, or that makes a more magnificent appearance. The calyx of the flower, when open, is nearly one foot in diameter, the inside, being of a splendid yellow color, appears like the rays of a bright star; the outside is of a dark brown. The petals, being of a pure white, contibute to the lustre; the vast number of recurved stamens in the centre make a fine appearance. Add to all this the strong, sweet fragrance and there is scarcely any plant which so much deserves cultivation. (Nicholson).

Cereus Michel′di. (6)

C. Pasacan′a, from Mexico. (7)

C. pugioniferus, having large and stout spines. (8)

C. Dumortier′i. (9)

C. Bauman'ni, from Peru.

C. Boxanen'sis, from Cuba.

C. coccin'eus, red ; from Mexico.

C. Donkelar'ii, from Brazil, where it grows among the Orchids.

C. gigant'eus, a large columnar Cactus, from Mexico.

C. Macdonal'diae, a handsome night-flowering species, from Honduras.

C. macroglyph'a.

C. Napoleon'is, from the Island of St. Helena.

C. nig'ricans, dark.

C. Olfer'sii, from Brazil.

C. tuberos'us, has tubers which the Mexicans steep in alcohol to use the extract as a remedy against rheumatism ; strange to say, it is not used internally, but rubbed on the affected part.

C. triangular'is, has traingular stems, from Mexico.

C. Havanen'sis cristat'a, a crested variety, from Havana.

Echinocereus is closely allied to Cereus, but has short stems armed with sharp, formidable spines.

Ech'inocereus eumeacan'thus (10)

E. pectinat'us, combed. (11)

E. viridiflor'a, green-flowered. (12)

E. chloran'thus, greenish.

E. dasyacan'thus, densily-spined.

Pilocereus. From *pilos* wool ; on account of the long white hair which resembles the hoary looks of an aged man. The best known species is

Pilocer'eus sen'ilis, Old Man Cactus. The resemblance of this Cactus to the top of an old man's head arrests the attention of many visitors. In Mexico, where this Cactus grows wild, it attains a height of 20 to 25 feet. (13)

Echinop'sis Eyries'ii, from *echinos*, hedgehog and *opsis*, like ; a small Mexican Cactus producing large, fragrant flowers.

Mammillar'ia, Nipple Cactus. The Cacti of this group consist of globular or cylinderical succulent plants, whose surface is not cut up into ridges, but is covered with many nipple-like tubercles, spirally arranged and ending in woolly cushions, from which bundles of spines emerge. The flowers are not large, generally purple, pink, white or yellow and issue from near the top of the plant. They are mostly Mexican species with a few from the West Indies and Brazil.

Mammillaria applanat'a, flattened. (14)
M. bi'color, two colored.
M. Bocassan'a.
M. can'dida. (15)
M. conoid'ea, cone-shaped.
M. crassispin'a elegans, thick-spined. (16)
M. elegans. (17)
M. fuscat'a.
M. Graham'mii.
M. Kramer'ii. (18)
M. micromer'is, small-flowered. (19)
M. min'ima, small.
M. montan'a, growing on mountains.
M. Nutal'lii.
M. Odierian'a.
M. pectinat'a, combed.
M. pusil'la, small. (20.)
M. rodan'tha, rose-flowered.
M. spinos'sissima, very spiny.
M. tuberculos'a, tubercled. (21)

Anhalonium, Spineless Cactus. This genus is closely related to Mammillaria; it consists of a few species not often met with in Cactus collections.

Anhalonium, fissurat'a, Living Rock. Looks indeed more like a fissured rock than a plant.

A. prismat'icum, "Seven Stars." A rare Cactus of pearl gray color. (22)

A. William'sii, Dumpling Cactus, might, from its peculiar shape, as well be named Cake Cactus.

Echinocactus, Hedgehog Cactus. This genus comprises many species of various forms; characteristic of all of them are the stout spines often produced in great numbers, 50,000 of them being found sometimes on a single plant. The spines of some species are used as toothpicks by the Mexicans.

Ech'inocactus bi'color, two-colored. (23)
E. capricorn'is, Goat's Horn Cactus.
E. corniger'eus, horn-bearer.
E. cylindrac'eus, cylindrical. (24)
E. electracan'thus, Lightning Thorn.
E. Em'ory.
E. Engelmann'i.
E. Gruson'ii. (25)
E. heloph'orus.
E. hexædroph'orus, Hexagon Cactus.
E. Lecon'tii.
E. longihamat'a, long-hooked. (26)
E. mul'tiplex, manifold.
E. myriostig'ma, many-dotted.
E. obvalat'us, fortified.
E. Orcut'ti, peculiar on account of its twisted ridges.
E. rhodophthal'amus. red-eyed. (27.)
E. saltillen'sis.
E. Simpson'ii.
E. Wisliczen'ii.

Phyllocactus, Leaf Cactus. These Cacti, of which there are only a few species, grow upon trees; like the Orchids, and have flat, leaf-like branches with notched margins. They have very handsome flowers, freely produced and are easily raised; for this reason they are quite popular.

Phyllocac'tus Ackerman'ni, King Cactus, is one of the handsomest species with rich crimson flowers.

Ph. latifrons, broad-stemmed ; has very large, creamy-white flowers.

Opuntia, Prickly Pear or Indian Fig Cactus, is a group of mostly North American species, quite distinct from those discussed thus far. They are fleshy shrubs with round, woody stems and many branches, generally much jointed. The flowers are not always showy; the fruit is pear or egg shaped, beset with prickles; those of O. vulgaris and Tune are edible and wholesome. In Mexico some Opuntias are largely cultivated for the rearing of the Cochineal insect which yields the genuine Carmine dye. The name of the genus is an old Latin name used by Pliny and is said to be derived from the City of Opus. (28.)

Opun'tia Dacuman'a elongat'a.

O. Em'ory.

O. Ficus Ind'ica, Indian Fig, from Mexico.

O. Tuna, West Indies.

Peres'kia, Barbados Gooseberry. From Nicolaus F. Peiresk of Air in the Provence, Senator and botanist. This genus is different from the others in having regular foliage leaves on the spiny branches, as seen in the illustration. (29.)

Pelecyphor'a aselliform'mis, Hatchet Cactus. From *pelekyphoros*, hatchet-bearing; so called from a fancied resemblance of the tubercles to a hatchet; *aselliformis*, woodlouse-like; this refers to the rows of scales which have been compared to the scaly back of a woodlouse.

Two plants not belonging here, but often found in Cactus collections on account of their resemblance to Cactus forms are mentioned here.

Gasteria punctata, from *gaster*, belly, referring to the swollen base of the flowers. This genus is allied to the Aloes; it has thick two-ranked leaves covered with many white scaly dots. From the Cape of Good Hope.

Rochea falcata, named in honor of M. de la Roche, a French botanist. A small fleshy South African plant of the Houseleek family (Crassulaceae).

R. C. PATTERSON.   N. PATTERSON.

PATTERSON BROTHERS,
FINE FLOWERS.
511 MARKET STREET AND 41 SIXTH AVENUE,
PITTSBURGH, PENN'A.

GROTTO AND SPRING. Schenley Park Fern House.

CURIOS.

PITCHER PLANT
Nepenthes Rafflesiana

Curios.

Every plant, and be it the commonest, simplest weed, yields interesting and instructive facts to the student; but there are among plants, as among people, "odd fish," with quite peculiar ways of their own, which make them especially interesting. To some of these, grown in the Conservatories, the attention of the visitor is called here.

PITCHER PLANTS.

Look at the handsome pitcher on the opposite page; if the illustration would give the natural colors, it would be light green, mottled with purple brown, a very pleasing combination. The pitcher grows from the apex of a leaf, connected with it by a long handle. The lid, which tightly closed the pitcher before the leaf had attained its full growth, now stands open and everybody is invited to drink. Such a cunning cup should offer a quite delicious beverage; when the writer examined the pitcher, still attached to the plant, it was half filled with a clear liquid; of course, it was at once tasted; to the disappointment of the investigator it was found to be tepid, insipid, with just a slight tinge of sweetness. The writer does not think it will ever be a popular drink hereabouts; but he hopes that to the insects, which are said to be attracted by a mysterious power to that fountain, it tastes like the very nectar of the gods, for it is the last drink these poor creatures get. In the Conservatory, insects do not get much encouragement; it is for this reason that there is seldom one found in the pitcher; but in Madagascar and other islands of the Indian Ocean where these Pitcher plants grow wild, they are said to be often filled with flies, ants, beetles and bugs who, having quaffed the kind nepenthe, forgot forever their earthly troubles.

Nepenthes, is the botanical name of the Pitcher plant and indicates its supposed properties; there are about thirty different species known, differing mainly in the size, shapes and coloring of the pitchers. The Conservatories possess at present the following:

Nepen'thes Dominiana, a garden hybrid with deep green pitchers, slightly spotted. [A. & S.]

N. Craig'ii. [A. & S.]

N. grac'ilis. [S.]

N. Hamilton'ia. [A. & S.]

N. Hook'eri. [S.]

N. hybrida, has dark-green pitchers 8 inches long, winged and fringed in front. [S.]

N. hybrida maculat'a, is a variety of the above with very long pitchers, streaked with reddish-purple upon a dark-green ground. [S.]

N. intermed'ia, pitchers 6 inches long by 2½ wide; swollen in the middle and having broad, fringed wings. [S.]

N. Mastersian'a, has pitchers 4½ inches long, 1¼ inches wide; deep claret-red. [A. & S.]

N. Morgan'iae, Mrs. Morgan's, has long flask-shaped pitchers, beautifully mottled with bright-red and pale-green when young, getting blood-red when mature. [A. & S.]

N. Rafflesian'a, named after Sir Stamford Raffles, has greenish-yellow pitchers with brown markings, very handsome. See illustration. [A. & S.]

More interesting even than those Asiatic Pitcher Plants are our North American Pitcher Plants or Side-Saddle plants. SARRACENIA, named after Dr. Sarrazin, of Quebec. These have various contrivances to allure the insect into the deadly cup, and others to make their exit impossible. They eminently deserve a place beside the Nepenthes and will no doubt get it as soon as good specimens can be procured.

Dros'era rotundifol'ia, Sundew. From *droseros*, dewy; *rotundifolia*, round-leaved. This plant is so tiny that it will be overlooked by many visitors; it has been discovered by one of the Assistants in unpacking plants which were wrapt in moss. It has a habit of growing between the soft swamp-moss (Sphagnum) and thus is quite likely to be found among the moss used for packing.

The specimens were planted in a flower pot with some of the Sphagnum and are doing nicely. Their tiny round leaves, scarcely more than half an inch in diameter, are bristling with hairs, each one having a minute, clear droplet at the end. If a small fly alights upon the leaf, it soon discovers that these drops are sticky and in its efforts to get away, becomes more and more entangled in those hairs; moreover, the hairs of the parts of the leaf not in contact with the insect, begin to bend over, approaching the poor creature and slowly, but surely enclosing and fastening it down on all sides. Here is the victim, entrapped, held fast as by an Octopus and doomed to be devoured or rather slowly digested. Darwin has made many and careful experiments with Drosera rotundifolia, and those who wish to know all about the tragedy, should read his book on Insectivorous Plants.

Mimos′a pud′ica, Sensitive Plant. From *mimos*, a mimic; *pudica*, chaste.

> A Sensitive-Plant in the garden grew,
> And the young winds fed it with silver dew
> And it opened its fan-like leaves to the light
> And closed them beneath the kisses of night.—*Shelley.*

Not only beneath the kisses of night will it close, but also beneath your kisses, if the attendants would permit you to try the experiment; such is the bashfulness of this pretty plant. Touch the leaves ever so slightly and they will close up, each little pair clasping tightly together until the whole double row of leaflets is thus folded up. If the whole plant is brushed over with the hand, all the leaflets close up; the four main divisions, of which each leaf is composed, draw near each other, the petiole droops down, the whole plant collapses. Does it possess nerves, capable of transmitting an outside impulse and causing a reflex action in the leaves and branches? Many have hinted at the posibility of such being the case, not only an account of the actions of the Sensitive plant, but also of the behaviour of the Venus Fly Trap and a number of other plants which act as if gifted with the sense of touch and even with will-power. But the closest microscopical examination has thus far failed to reveal any organs comparable to our nerves. Mimosa pudica is a native of tropical America; many other plants of the same family (Leguminosae), also some Oxalis, share the property of being sensitive to touch, but in a less degree. [A. & S.]

Desmod′ium gir′ans, The Telegraph Plant. From *desmos*, a band, alluding to the fact that the stamens are grown together by their filaments, an arrangement found, however, in many other flowers of the Pulse family (Leguminosae); *gyrans*, moving. The name Telegraph plant or Signal plant it owes to the following pecularity:—Each leaf consists of three leaflets, a large one, between two and three inches long and about one-fourth as broad, of illiptical shape, and two quite small leaflets, of the same shape and near the base of the large one. These small leaflets move continually up and down, either with a gradual, almost insensible movement or in jerks; sometimes both leaflets are raised or both down, sometimes one is up the other down, somewhat in the manner of the signals (Semaphores) on the railroads. In these movements the tips of the leaflets describe an elliptical orbit and it takes from half a minute to one minute and more to go through one rotation, the rate depending upon the temperature and other conditions of the atmosphere. The large leaflets do not go through this motion, but every evening they droop down as if going to sleep ; at sunrise they rise up again, taking an almost horizontal position. The plant is a native of the East Indies.

[s.]

Pil'ea microphyl'la, Artillery or Pistol Plant. *Pilos*, a cap, from the shape of one of the parts of the perianth ; *microphylla*, small-leaved. This plant of the Nettle family (Urticaceae), will remind the reader of a Lycopodium. It is a rather insignificant, shrubby weed with inconspicuous flowers, but it has a peculiarity to which it owes its common name ; at blooming time the branchlets are covered with many little purplish buds; when bright sunlight falls on these, one after another begins to open with a slight puff, discharging a cloudlet of dust. These little guns are the staminate flowers; four stamens are coiled up in these, impatiently waiting for their liberation ; the rays, or the heat they carry, seem to unlock the calyx, the stamens uncurl with a jerk, pushing back the sepals and discharging the pollen into the air. It is a pretty sight to see this canonade going on, which is especially lively if the plants have been watered shortly before. After the battle is over, the branchlets are covered with tiny white crosses—the stamens with their emptied anthers.

Marcgrav'ia paradox'a, the Shingle Plant. This plant seems determined to play the part of vegetable shingles, its leaves are round thickish, rather large, and they grow closely upon the rock, wall, trunk or whatever object forms their support ; besides, they overlap each other shingle fashion, so that they form a cover, perfectly impenetrable, at least to rain.

These are some of the curios which interest the scientists, puzzle the people, and make thoughtful persons think. Various other plants, mentioned in former chapters, might find a place here.

The Ferns.

Maidenhair Ferns. Schenley Park Conservatory.

FERNS.

What makes Ferns so popular? What is there to be admired in these plants, entirely devoid of showy or pretty blossoms? No doubt it is the variety and beautiful symmetry of their patterns, the graceful manner in which their erect or drooping fronds are borne. A shady ravine, a rocky glen, filled with a profusion of Ferns, is it not a pleasure to look upon? Did you ever notice how the bunches of Christmas Ferns, peeping out from almost every crack and crevice along the little stream in Panther Hollow add to make that winding valley so picturesque? What could take the place of the ferns in airy bouquets or in your conservatory to form a soft and pleasing background to the flowers?

Ferns are among the highest forms of the great series of flowerless plants, of which the mosses, liverworts, seaweeds and toad-stools, mould and mildew, yeast and bacteria belong. The botanist uses special terms for the different parts of a fern. The underground stem, from which most ferns grow from year to year is called a *rhizome*; the stalks growing from the rhizomes are *stipes*, they support the *frond*, which answers to the leaf of other plants. The frond consists sometimes of one single, entire blade, as in the Nest-fern.; but more frequently it is composed of smaller blades called *pinnae*, meaning feathers; these may again be sub-divided into smaller leaflets called *pinnules*. The veins of the ferns are peculiar and quite different from the parrallel or netted veins of the leaves of flowering plants; in most ferns the veins are forked, in some they form symmetrically arranged meshes with branching veinlets inside of them.

The fruit in most ferns is produced on the under side of the fronds; it is disposed in clusters, either round, or oval, or narrow, near the middle or near the margin; these clusters are called *sori*, and they are often covered with a protective scale called *indusium*, which withers away as the fruit ripens. Your magnifying glass reveals the fact that the sori are heaps of peculiar brown, roundish bodies; these are the *sporangia* or spore cases which burst open when ripe, and discharge an exceedingly fine powder, the *spores*, which are the seeds of ferns and other flowerless plants. Examined under a good microscope, the spores appear as pretty spherical or oval bodies, often with characteristic markings and resembling much the pollen which is discharged from the stamens of flowers. Unlike the seeds of the flowering plants, the spores contain no embryo, or little plantlet.

The study of ferns is very fascinating, and it is not difficult at all to become acquainted with the names and peculiarities of our native ferns. If the fair reader, repeatedly appealed to, is interested in the study of nature, she should organize a "Fern Club" among her friends for the purpose of collecting and studying all the ferns which are growing wild in the neighborhood of Pittsburg and Allegheny. Gray's Manual of Botany and Robinson's "Ferns in their Homes and Ours" will give you all the help you need; and if you faithfully carry out your programme, you will derive more information and pleasure than you now imagine.

Ferns are classified according to the peculiarities of their fruiting arrangements; the shape of the sori mentioned above, the place where they are attached, the presence and nature of the indusium and other points have to be observed in order to know to which group or genus a fern belongs. In giving the extensive list of ferns cultivated in the Conservatories, let us begin with the most popular and best known of all, the

MAIDENHAIR FERNS.

Adiantum is the latin name of this genus; it means *unwetted*, from the fact that its smooth foliage repels the rain drops. The Adiantums are exceedingly graceful on account of their multi-compound fronds, the pretty shape of the leaflets which are generally drooping, being borne on almost hair-like, shining stemlets, while the stipes are mostly dark and polished. The fruit dots or sori are formed at the outer margin of the pinnules, and the end of the scolloped edge is turned over, thus forming the protective indusium. Our native growing Maidenhair Fern, Adiantum pedatum, is one of the prettiest species, and much cultivated in Europe; but the principal home of these ferns are the tropical countries and islands of the Pacific; there they abound in hundreds of different species. The pleasing fern group on page 178 represents about the middle of the foreground. Adiantum Farleyense, the Queen of the Maidenhair Ferns; it is a most fair queen with the softest of tresses. To the right below is

Adiantum gracillimum, the most graceful, with the tiniest of leaflets; above we see Adiantum cuneatum with wedge-shaped leaflets and other species. The following is the list of all the species of Maidenhair Fern grown at present at the Conservatories:

Adian'tum Aneiten'se, from the Aneiteum Isles.	[s.]
A. Bal'lii.	[s.]
A. bel'lum, handsome; from Bermuda.	[s.]
A. ciliat'um, hairy.	[s.]
A. concin'um, neat; from tropical America.	[A.]
A. Craig'ii,	[s.]
A. cuneat'um, wedge-shaped; from Brazil.	[A. & S.]
A. cuneat'um grandiceps.	[A. & S.]
A. cuneatum variegatum	[s.]
A. decor'um, decorous.	[s.]
A. Dreer'ii.	[s.]
A. Farleyens'e, a variety of A. tenerum, tender.	[A. & S.]
A. formos'um, beautiful; from Australia.	[A. & S.]
A. gracil'limum, most graceful; of garden origin.	[A. & S.]
A. Lathom'ii.	[S]
A. macrophyllum, long-leaved; from tropical America.	[S]
A. puesb'cens, downy.	[s.]
A. Santæ Catharinæ, a variety of	[A. & S.]
A. trapezifor'me, rhomb-leaved; from the West Indies.	[A. & S.]
A. Victoria, named after Queen Victoria.	[A. & S.]
A. William'sii, this is one of the most beautiful species; of greenhouse origin.	[A. & S.]

EAGLE FERNS.

Pteris is the name of an extensive genus, the best known species of which is certainly Pteris aquilina, the Eagle Fern, Brake or Bracken which grows so abundantly in this country, in England and the European continent and in many other parts of the world.

> The heath this night must be my bed;
> The bracken curtain for my head;
> My lullaby the warders tread
> Far, far from love and thee, Mary.
> —*Lady of the Lake.*

The name Pteris is derived from *pteron*, a wing, to which the large and broad fronds of the Eagle Fern are compared. These ferns, instead of being soft and pliant like the Maidenhair Ferns, have generally rigid fronds, and the sori are continuous and follow the margins of the pinnae and their lobes. The examination of a few of the species named here will enable the fern student to recognize other Eagle Ferns at sight.

Pter'is argyr'æa, silver-banded, is a form of P. quadriaurita, four-eared, with a band of white down the centre of the frond.
[A. & S.]

P. cret'ica, from Crete, has the frond divided into long and narrow pinnae; grows wild in Florida and other semi-tropical countries.

P. cret'ica albo-lineata, white-lined, is a variety often found in greenhouses. [A. & S.]

P. hastat'a, halberd-shaped; from Africa. [S.]

P. leptophyl'la, slender-leaved; from Brazil. [S.]

P. longifol'ia, long-leaved; from the West Indies. [A.]

P. Owrar'dii, a garden variety. [S.]

P. palmat'a, hand-shaped; from tropical America. [A.]

P. serulat'a, saw edged; from India. [A. & S.]

From this fine species several varieties are derived, as:

P. serulata cristata, crested. [A.]

P. serulata magnifica. [S.]

P. serulata nobilis. [S.]

P. trem'ula, trembling; from Australia. [A. &. S.]

P. trem'ula Smithiana. [S.]

P. Veitchii, [A.]

P. Victoria, [S.]

P. Mayii, are garden varieties. [S.]

TREE FERNS.

There is probably no public Conservatory which boasts of a finer collection of Tree Ferns than Schenley Park Conservatory; the domed pavilion devoted to it is one of the most charming, refreshing and interesting parts of the great establishment; it gives a very impressive picture of a fern forest. The tall tree ferns with mossy trunks are elegant specimens of

TREE FERNS.
Dicksonia antarctica.
Schenley Park Conservatory.

Dickson'ia antarc'tica, the genus is named after James Dickson, a famous British botanist ; it contains about 30 species, half of which are tree-like ; one of the handsomest ferns of our woods, sometimes called Sweet Fern on account of the fragrance of the fronds, noticed especially when they are wilting, belongs to this genus (Dicksonia pilosiuscula). The sori are round and placed at the end of the veinlets, near, but not at the margins, and they are covered with either cup-shaped or two-valved indusiums. The species *antactica*, is a native of East Australia and Van Dieman's Land ; it attains a height of 30 to 35 feet. [A. & S.]

The following are related to the Dicksonias :

Cibot'ium (Dicksonia) **Bar'ometz,** a tree fern from China and the Malayan Peninsula. The name is derived from *cibotion*, a small box ; those who examine the indusiums of these ferns will readily understand what suggested this name. [S.]

Cibotium regale, (Dicksonia regalis) from Mexico. The handsome fronds of this fern can be seen in the illustration on page 140. [S.]

Cibotium (Dicksonia) **Shied'ei,** is also a Mexican species, growing to the height of 10 to 15 feet. [S.]

Sitilobium cicutar'ium (Dicksonia cicutaria). From *citos*, wheat and *lobos*, a lobe, referring to the shape of the lobes of the fronds. Does not grow tree-like, but has a creeping rootstalk; native of Mexico and the West Indies. [A. & S.]

Dennstaedtia (Dicksonia) **davallioides,** Davallia-like ; grows also from an underground stem and has fronds of thinner texture than the above. From Australia. [S.]

D. Young'ii. [S.]

Alsoph'ila australis, from *alsos*, a grove and *philo*, to love, in reference to the favorite location of these ferns. A very fine form of tree ferns from New Holland, of which there are numerous other species ; they are distinguished from the Dicksonias in having the sori either on a vein or in the forking of a vein and not being provided with an indusiums. [A. & S.]

Cyathe'a dealbat'a, from *kyatheron*, a little cup, in reference to the appearance of the sori on the back of the fronds ; *dealbata*, whitened ; the fronds being white on the under side ; from New Zealand. This is also a representative of a genus of magnificent

tree ferns; the sori are placed as in Alsophylla but covered by an indusium. [s.]
 Cyath'ea medullaris, pithy; from New Zealand. [s.]
 Tod'ea Barbara, named in honor of Henry Julius Tode of Mecklenburg. A handsome small tree-fern from Australia. [s.]

 Gleichen'ia longipinnat'a, named in honor of the German botanist Dr. W. F. Gleichen. A very graceful and peculiar fern, the fronds of which have a habit of repeatedly dividing into two forks and the pinnules of which are quite minute; there are some fine specimens in the fern house. [s.]
 G. Spelun'cæ, Cavern Fern; and
 G. semivestit'a, half-clothed, are synonyms for
 G. circinat'a, rolled up; from Australia. [s.]
 Davallia, this genus is named after E. Davall, a Swiss botanist; it has generally very finely divided fronds and the sori are at or near the margin, the indusium is on the ends of the veins and open at the apex.
 Daval'lia affin'is, related; from Ceylon. [s.]
 D. bullat'a, blistered-leaved; from the East Indies. [A. & s.]
 D. fijen'sis, from the Fiji Islands.
  Variety plumosa. [s.]
  Variety major. [A.]
 D. Moorean'a, a synonym of D. pallida; from Borneo.
                   [A. & s.]
 D. par'vula, little; from Borneo. [s.]
 D. pectinat'a, combed; Polynesian Islands. [A.]
 D. Tyerman'ni, from the West Coast of Africa. [A. & s.]
 D. (Microlepis) hirta, hairy; from North India. [s.]
 The variety cristata has crested pinnae. [A. & s]
 Hypolep'is repens, from *hypo* under, and *lepis* a scale; alluding to the covering of the sporangia; repens, creeping; from tropical America. [A.]
 Cheilan'thes hirta, Lip Fern. From *cheilos* lip, and *anthos* flower; the sori are close to the margin and are covered by the reflexed edge of the lobes. Rather small ferns, generally very hairy; the above is from the Cape of Good Hope. [s.]

Himalaya Hanging Fern (page 187) in the middle; below, Latania aurea (page 26); above, frond of Phœnix spinosa (page 34); to the right, Corypha australis (page 25) and Cycas revoluta (page 40).
Schenley Park Conservatory.

Onych'ium aurat'um, from *onyx*, onichos a claw, alluding to the shape of the lobes of the fronds; *auratum* golden; from Japan.
[A. & S.]

Lomar'ia gib'ba; from *loma* an edge; *gibba* gibbous. The sori in these ferns are long and narrow, occupying the space between the midrib and the margin; the indusium is formed by the edge of the frond being turned inward. From New Caledonia. [A. & S.]

L. (Stenochlæna) **scandens,** also named

Acrostichum scandens, a climbing species from the Himalaya mountains. [A.]

Blech'num Brazilien'sis, from *blechnon,* the Greek name of a fern. Resembles the Lomaria, having also the sori parallel with the midrib, but close to it and the indusium is separate from the edge of the frond. [S]

Woodwar'dia rad'icans, Chain Fern. Named after the English botanist, Thomas Jenkinson Woodward; *radicans* rooting. The name of Chain Fern is suggested by the arrangement of the sori which run parallel and close to the midveins. From Europe. [S.]

POLYPODY FERNS.

These ferns have round sori not covered by an indusium ; the name is derived from *polys,* many and *podion,* a little foot, on account of the appearance of the root-stalk and its appendages. One of the species, Polypodium vulgare, the common Polypody, is quite abundant on rocks in the Allegheny mountains.

Polypod'ium aur'eum, golden, is found in Florida and the warm countries. [A.]

Polypodium sporadocar'pum, spore fruited; is a variety of P aureum. [S.]

P. Phymatod'es, from the East Indies and the Polynesian Islands, is largely cultivated. [S.]

Polypodium (Goniophlebium) sub-auriculatum, the Himalaya Hanging Fern. This most graceful Fern is shown in the illustration on page 186.

SPLEENWORTS.

Asplenium, from *a* not, and *splen* spleen; at the time when old women and herb doctors were consulted principally in cases of sickness, the Spleenwort was thought to be a great remedy for certain complaints. There are several Spleenworts among our native ferns; they are characterized by an elongated sorus, being covered by an indusium which is attached to a vein and opens toward the midrib.

Asplenium Belan'geri, from the Malay Peninsula. [A. & S.]

A. bulbif'erum, bulb-bearing; having little bulbs upon the upper part of the fronds from which new plants develop. From New Zealand. [S.]

A. bulbif'erum Wollaston'ii. [S.]

A. formos'um, beautiful; from tropical America. [S.]

A. (Thamnopteris) Nidus avis, Bird's Nest Fern. These are grand ferns with immense undivided fronds growing in a circle, the young unrolling fronds in the middle. Might form a convenient nest for a huge bird. Notice the long and narrow sori on the back of the fertile fronds, forming parallel lines closely together. The specimens in Schenley Park Fernhouse are especially fine.

[A. & S.]

Shield Ferns. These ferns have round sori covered with a round or kidney-shaped indusium; some of the most common ferns of our woods belong to this genus. Aspidium, from *aspidion*, a little buckle, is the name of the genus proper.

Aspidium (Polystichum) angular'e, the Soft Shield Fern. A garden variety of A. aculeatum. [A.]

Closely related are:

Lastrea aristat'a, (Nephrodium aristatum). From *nephros*, a kidney, referring to the shape of the indusium ; *aristatum*, awned. From the Philippines. [A.]

L. patens cristat'a, crested. [A.]

L. crispum, curled. [A.]

Sword Ferns. These are so called on account of the long and narrow fronds which characterize most of these ferns ; the pinnae are so crowded that from the distance the frond looks like a sword-shaped leaf. The Latin name NEPHROLEPIS, is derived from *nephros*, a kidney, and *lepis*, a scale, referring to the shape of the indusium.

STAGHORN FERN,
(See page 190)
Schenley Park Fern House.

Nephrolep'is Davallioid'es fur'cans, Davallia-like and forked; the pinnae dividing in two and more parts at the apex. [A. & S.]

N. Duffii, an elegant species from Duke of Yorks island. [A.]

N. exaltat'a, lofty, grows wild in Florida; very largely cultivated. [A. & S.]

N. pectinat'a, comb-like; a variety of N. cordifolia. [s.]

N. plum'a, feather; from Madagascar. [s.]

N. rufes'cens tripinnatifida, reddish, three-pinnatified. [A.]

N. Zollingerian'a. [s.]

Staghorn Fern. These ferns are exceedingly odd with their long and leathery fronds, looking like a forked tongue rather than like a stag or elk-horn. They are air plants and grow upon trees which they in time encircle with the large kidney-shaped scales from which the fronds grow. The fruit forms in large patches on the under side towards the apex of the fronds. The scientific name PLATYCERIUM, comes from *platys*, broad and *keros*, a horn.

Platycer'ium alcicor'ne, elk-horn; is the species of which the Schenley Park Fern house possesses so fine and valuable specimens. They are from temperate Australia. [s.]

There remains still to be mentioned the SELAGINELLAS, flowerless plants of the Lycopod family, allied to the ferns, but more moss-like in appearance. They are mostly from tropical America.

Selaginella caes'ia, gray [A.]

S. caes'ia arbor'ea, tree-like. [A.]

S. caules'cens, stemmed. [A.]

S. denticulat'a. small-toothed. [A. & S.]

S. Emilian'a. [s.]

S. Kraussian'a aur'ea, golden. [s.]

S. Poul'teri. [s.]

Closing Remarks.

E. M. BIGELOW,
DIRECTOR OF PUBLIC WORKS.
Pittsburg, Pa.

THE PARKS.

Cities owning fine parks are justly proud of them; parks are blessings in more than one way; they are fresh air reservoirs, the salubrious influence of which extends over a large area beyond the park limits; they give opportunity for healthful outdoor exercise, so necessary for city people; they are the children's paradise, they are the Mecca for people with elegant turnouts as well as for the modest citizen and loving father who gives his family an outing, on Sunday, in his grocery wagon. Pittsburg has been fortunate in becoming possessor of park sites which in situation and natural advantages cannot be easily matched. To make proper use of these advantages, to blend nature and art so as to obtain the most favorable results, is the problem now on hand. That it will be well solved the citizens of Pittsburg are confident; they know that the present chief of public works, Mr. Bigelow, whose well known features are represented on the opposite page, has the ambition to make of smoky Pittsburg, a fair, a beautiful city and that especially the parks are near his heart, and it will not be his fault, if they do not become equal to the most famous parks in the country.

The people of Pittsburg greatly appreciate Mr. Bigelow's energetic efforts in behalf of the Parks and already there is a movement on foot to commemorate in a befitting manner the director's invaluable services in creating out of a wilderness and presenting to the public, in so short a time, a royal pleasure ground for Pittsburg's inhabitants of all classes and all ages.

SCHENLEY PARK covers 431 acres, 412 of which have been given to the city by Mrs. Schenley, and 19 acres have been bought. The natural relief of this large tract, its deep gulches, its undulating hills, its points of magnificent outlook, are great advantages for a park at the start. The improvements thus far accomplished are the completion of 6½ miles of driving roads to which as many more miles will be added; a lovely promenade through Panther Hollow, crossing many rustic bridges and leading through some picturesque spot at every turn; other shaded walks have been projected, one of them leading near the carriage road so that the pedestrians may enjoy the sight of the great *Corso*.

The lake at the entrance to Panther Hollow is being deepened and widened, and the surrounding scenery will be greatly beautified; there will be pleasure boats in summer, skating in winter. A bridle path to Mt. Airy, and also walks to and through that region below the Race Tracks will open up an entirely new region; it will be made of easy access through a bridge of graceful structure thrown across Panther Hollow. The Zoo, too, will present in time a more dignified appearance; the animals will receive quarters in keeping with their natural tastes, and their number and variety will be vastly increased. Lovely groves and shaded avenues will afford welcome protection against the too energetic rays of the sun which now are scorching with impunity the bare hilltops, and those who venture upon them. Time and money are the only things needed to carry out these and other intended improvements.

HIGHLAND PARK, also, is unique in its location, covering an area of about 340 acres, surrounding the large reservoir with terraces, slopes, depressions with little lakes, elevations with most charming views up and down the Allegheny Valley, and other natural features which can be utilized to produce charming details. When the superb plan, designed by Mr. Berthold Frosch, Director Bigelow's landscape gardener and engineer, will be carried out, Highland Park will be a gem among the parks of the country, and a place which every Pittsburger will take pride in showing his friends from abroad.

HERRON HILL PARK, beautifying the surroundings of Herron Hill reservoir, the highest elevation in the county, is greatly enjoyed by the inhabitants of the densely populated quarters of the "Hill" district.

ALLEGHENY PARK. Allegheny boasted of its park before Pittsburg ever dreamed of possessing such a luxury; indeed it may be said that the park was there before the City of Allegheny existed. As early as 1787 about 100 acres of land, surrounding what was then the town of Allegheny, were given by act of Legislature as common pasture land to said town. In 1869 the commons became a public park. There is probably no other city in the country which has a park so situated that nearly every citizen has a piece of it at his very door or within a few minutes walk; and with its well patronized playgrounds, its umbrous promenades, its well kept flower beds, its picturesque lake and its fine monuments, it has an air of maturity, of matter of-fact existence, which the Pittsburg parks will not be able to assume for some time.

RIVERSIDE PARK, Allegheny's most recent acquisition, is a tract of 200 acres of wood and farm land stretching from Perrysville Avenue west towards Woods Run, forming mainly what used to be Watson's farm ; it was purchased by liberal spirited citizens and presented to the City of Allegheny. The official presentation and acception took place on the 4th of July, 1894 with appropriate ceremonies. The picturesque glen which forms the most attractive part of the tract, has long been a favorite picnic ground and an eldorado for botanists, who found there not only a great variety of native trees, but several species of herbs not found in other localities of this vicinity. Nature needs only a little assistance to make of this tract a very pleasant outing ground. The highest part of the park is a round knoll near the Perrysville avenue entrance ; from there the visitor enjoys a magnificent view of stretches of the Ohio river and its shores and of other parts of the surrounding country. This knoll will in all probability be the future site of the Allegheny Observatory; plans are being prepared for the enlargement and refitting of this excellent institution, thanks to the munificence of a well known citizen who offered to pay the expenses.

Superintendent Wm. Hamilton has the active support of Mr. Robert McAffee, Chief of Public Works in Allegheny, in all matters pertaining to the improvements of the parks, while Mayor Wm. M. Kennedy, himself a lover of nature and an experienced botanist, takes the deepest interest in the future of the parks, and gives valuable suggestions. He proposes, among other things, to create in the new park an Arboretum in which all the native American trees and shrubs which can be made to grow in this climate, shall be represented by one or several specimens and labeled so as to give information in regard to their name order, and use. No doubt the people of Pittsburg and Allegheny can look forward with great expectations to the future of their parks.

PROPOSED EXTENSION OF SCHENLEY PARK CONSERVATORY.

As may be seen from the sketch on the next page, an extension to this large Conservatory is already proposed and recommended by its managers. It consists of the addition of three wings to the Conservatory Building proper and a number of houses of simpler construction. The new wings are to be used for the raising of specialties, such as Orchids, Cacti and other tropical plants which

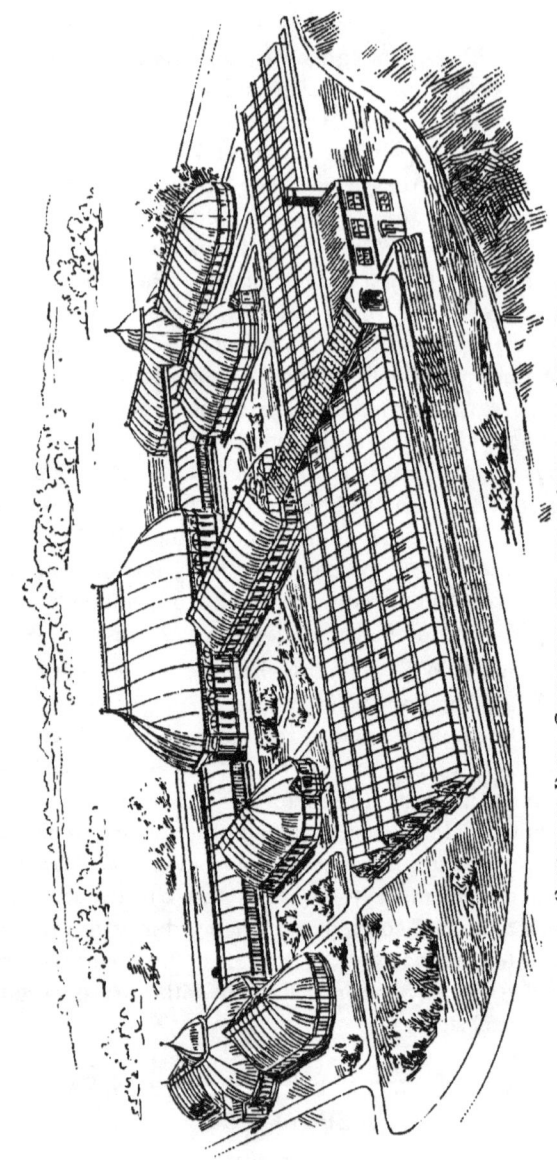

Schenley Park Conservatory, with Proposed Extensions.

need special attention, while the other buildings are to be the winter quarters for the many thousands of bedding and carpet plants which adorn the grounds during the summer season. This will enable Supt. Bennett to devote the whole of the main building to purposes of a Public Conservatory, and to give proper room to the many new additions which are continually arriving. The plans are, however, subject to considerable change, and their execution will mainly depend upon the time and measure in which the funds become available for the purpose.

BOTANICAL GARDENS.

As soon as the promised labeling of all the plants in the Conservatories will be carried out and completed, these institutions will fill to a great extent the purposes of botanical gardens which form such a conspicuous educational feature in all large and many small European cities ; without such an arrangement, the purpose of this Guide will only be partly accomplished. It is, however, reported that Director Bigelow is contemplating plans tending to the establishment, at some future time, of a botanical garden approaching the best institutions of a similar nature existing and comprising the prominent features of these, viz :

The various plant forms and families represented by living and characteristic specimens systematically arranged.

Groups of plants illustrating the character of vegetation in the various geographical plant regions.

A section for the cultivation of all plants useful in the arts and industries, as well as the food and medicinal plants of the world.

A botanical laboratory for the carrying on of investigation and experiments.

The proximity of that grand institution, the Carnegie Library, now nearing its completion, will make it possible for anyone to get information on the subject from the world's literature on botany ; a most rare opportunity for the student ; but even without books the young people will learn more botany by observation and absorption in these botanical gardens, than they are able now to get in the schools.

The Botanical Gardens will be one of the institutions that will mark Pittsburg in the future as a center of learning as it is now a center of manufacture and industry.

THE FLORISTS.

The establishments of the florists of Pittsburg and Allegheny, some of them quite extensive, have always been open to the public; they are conservatories on a smaller scale, though the variety of plants raised there is often quite large and the visitor has the advantage that he can buy and take home any specimen he desires.

The opening of the Phipps Conservatories has not diminished the demand for flowers, but has materially increased it. This fact proves that these institutions greatly help to develop an interest in flower culture, a love for the beautiful. But the customers of the florists are becoming more discriminating in their choice and ask for many kinds of plants now which were never in demand before. Florists are obliged to make many additions to the lists of flowers in their stock; enterprising as they are, they anticipate the public's more descerning and exacting demands and the consequence is now a varied display of splendid novelties in the windows of their town stores, beautiful to look upon and far superior to what had been offered in former years. Thus the Conservatories have indirectly contributed to improve floriculture in these cities in quantity and quality.

How intimately is the florist's business connected with our lives in joy and sorrow! From birth until death, in all important events we celebrate, of public nature, in church or in the private family circle, he must assist with his floral treasures that the scene may be befitting the occasion.

It is a beautiful calling, but an arduous one and one that requires experience, judgment, tact and taste.

THE GUIDE.

As the author's work upon this book is done and the compositors have wrestled with the Greek and Latin words and have been victorious, a few words may be said about the technical and artistic features of this book.

The photographic views of plants and interiors in the Conservatories have all been taken by the author; most of them have been engraved by the FORT PITT ENGRAVING CO., some by the ANDERSON-HOTZ ENGRAVING CO. Many of them will be conceded to be works of art equal to the best work in this line done in the country.

A few of the pictures had appeared first in the *American Florist* and in the "Proceedings of the Academy of Science and Art in Pittsburg," illustrating a paper "A Walk Through Allegheny Conservatory," read before said society by the writer of this guide. A few of the smaller illustrations were first published in the *Pittsburg Bulletin.*

The typographical and press work was done in the establishment of FOSTER, DICK & CO.; these gentlemen deserve the author's thanks for their pains-taking work. This Guide is the first book printed on their new four-roller Campbell cylinder press.

Thus this book throughout represents Pittsburg enterprise and it is herewith dedicated to the people of Pittsburg and Allegheny.

CORRECTIONS.

PAGE	20,	line 7,	Latania instead of	Lantania.
"	21,	" 37,	Macrozamia, "	Zamia.
"	31,	" .25,	pinnae "	prinnae.
"	54,		Sanchezia "	Sanchizia.
"	54,		Strobilanthus "	Strabilanthus.
"	54,		Maranta "	Marantha.
"	81,	last line,	Dendrobium "	Dendrobiom.
"	82,	line 33,	corolla "	calyx.
"	91,	" 21,	verrucosa "	vrerucosa.
"	104,	" 17,	Hydrangea "	Hydrangia.
"	131,	" 36.	Centradenia rosea.	
"	133,	" 17,	mimo "	mimno.
"	134,	" 6,	Dr. Alexander Garden.	
"	134.	" 10.	Hoffmannia instead of	Hoffmania
"	135,	" 14,	at the apex "	of the apex.
"	142,	" 30,	Antigonon, "	Antigonous.
"	151,	" 19,	tropic "	topic.
"	159,	" 2,	Eichhornea (Pontederia) crassipes.	
"	164,	" 15,	verified instead of	varified.
"	181,	" 19,	pubescens "	puebscens.
"	182,	" 20-24,	serrulata "	serulata

INDEX

OF THE GENERIC AND COMMON NAMES OF THE PLANTS MENTIONED IN THE GUIDE.

Aaron's Beard	132	Aspidium	188
Abutilon	130	Asplenium	188
Abyssinian Banana	20, 38	Astrocaryum	29, 30
Acacia	132	Attalea	30
Acalypha	131	Azalia	104, 108
Acanthophœnix	29	B.l-am	130
Acanthus	69	Bamboo	21, 44
Achimenes	104, 118	Bambusa	44
Achyranthes	135	Banana	21, 37, 59
Acorus	58	Backeria	88
Acrostichum	187	Bead Tree	130
Ada	91	Beaucarnea	44
Adiantum	178, 180	Bedding Plants	134
Æchmea	67	Begonia	104, 120, 126
Aerides	91	Bignonia	147
Agapanthus	110	Billbergia	66
Agave	43	Bird of Paradise Plant	39
Aglaonema	58	Birthroot	143
Air Plant	67	Blackburn's Sabal Palm	7
Alisma	159	Blechnum	187
Allamanda	146	Bletia	88
Alocasia	57	Borassus	24
Aloe	45	Bougainvillia,	142
Alsophila	184	Bourbon Palms	26
Alternanthera	134	Bowstring Hemp	47
Amaryllis	104, 111	Bracken	181
Amomum	68	Brahea Glauca	24
Ananassa	66	Brake	181
Angraecum	91	Brassavola	84
Anguloa	82, 91	Brassia	91
Anthericum	110	Broom Palm	27
Anthurium	54, 56	Broughtonia	88
Antigonon	142	Brunfelsia	133
Aponogeton	159	Burlingtonia	1
Aquatic Plants	149-160	Cabbage Palm	20, 21, 29
Aralia	31	Cabomba	158
Araucaria	20, 42	Cacti	161-169
Ardisia	132	Caladium	54, 57
Areca	20, 28, 29	Calamus	30
Arenga	29	Calanthe	91
Aristolochia	143	Calathea	61
Arrowroot	61	Calceolaria	104, 112, 113
Artillery Plant	176	Calla	55, 58
Arum Family	55	Camellia	129
Arundo	44	Camphylobotrys	134
Asparagus	64	Canna	60
Aspidistra	111	Cape Jessamine	134

Cape Pond Weed	159	Cypripedium	82, 83, 96, 98, 114
Cap Orchid	94	Cyrtopodium	92
Cardamom	68	Cytisus	112
Carludovica	21, 36	Dactylis	65
Caryota	20, 28, 30	Dæmonorops	32
Catasetnm	91	Dasylirion	44
Cattleya	82, 85, 86	Date Palm	20, 21, 33
Cedrus	41	Davallia	185
Centaurea	135	Dendrobium	81, 84, 87
Centradenia	131	Dennstaedtia	184
Century Plant	43	Deodar	41
Cereus	165	Desmodium	175
Ceroxylum	31	Dichorisandra	66
Cestrum	145	Dicksonia	183, 184
Chain Fern	187	Dieffenbachia	21, 54, 55
Chamædorea	31	Doryanthes	111
Chamaerops	20, 24, 25	Dove Flower	94
Cheilanthes	185	Dracæna	21, 45, 54, 63
Chili Pine	41	Dragon's Blood	32
Chysis	89	Dragon Tree	21, 63
Christmas Fern	179	Drosera	174
Chrysanthemums	105, 127, 129	Dumb Cane	6
Cibotium	184	Dumpling Cactus	107
Cineraria	102, 104, 107	Duranta	133
Cissus	144	Dutchman's Pipe	143
Clerodendron	21, 22, 147	Eagle Fern	181
Climbing Poke	143	Easter Lily	104
Club Rush	160	Echeveria	136
Cobæa	146	Echinocactus	167
Coccoloba	20, 44, 129	Echinocereus	166
Cochlioda	91	Echinopsis	166
Cocks-Foot Grass	65	Eichhornea	159
Cocoanut Palm	21, 31, 77	Elais Guineensis	23
Cocos	31	Elephant's Ears	57
Codiæum	71, 72	Elkhorn Fern	189
Coelogyne	85	Epidendrum	81
Coffee Tree	133	Eranthemum	70
Coleus	134	Eucalyptus	45
Colocasia	57	Eucharis	111
Conifers	40	Eugenia	131
Coral Plant	67	Euonymus	45
Corn Flower	135	Eupatorium	134
Coryanthes	92	Euphorbia	130
Corypha	25	Euryale	155
Costus	68	Fan Palms	20, 24-27
Cotyledon	136	Feather Palms	29-34
Crane's Bill	107	Ferns	177
Creeping Sailor	132	Ficus	21, 46
Crinum	112	Fig Marigold	136
Croton	54, 71-75	Fisherman's Basket	112
Crown of Thorns	130	Fittonia	69
Cupania	45	Flamingo Plant	57
Curculigo	21, 69	Florida Moss	67
Curios	171-176	Flowering Maple	130
Curmeria	78	Flowers	101-186
Cyanophyllum	71	Foliage Plants	55-75
Cyathea	184	Fuchsia	104, 119
Cycads	20, 21, 22, 40	Franciscea	133
Cyclamen	99, 103, 105	Freesia	104, 108
Cymbidium	92	French Daisy	129
Cyperus	160	Gardenia	134

Gasteria	169	Leaf Cactus	168
Gastonia	46	Ledenbergia	143
Genista	104, 112	Lemnocharis	159
Geonoma	32	Licuala	26
Geranium	104, 107	Lily	104, 110
Gesnera	105, 118	Limatodis	91
Giant Arrowhead	159	Linum	130
Ginger	68	Lip Fern	185
Ginseng	131	Little Coco	21
Gleichenia	185	Living Rock	167
Gloriosa	142	Livistonia	26
Gloxinia	104, 118	Lobelia	135
Goniophlebium	21, 187	Lomaria	187
Goodyera	95	Lotus	155, 158
Grasses	65	Lycaste	82, 87, 92
Grevillea	70	Macrozamia	20, 21, 41, 42
Ground Ivy	71	Maidenhair Fern	178-181
Ground Rattan Palm	27	Mammilaria	166
Gum Tree	45	Manettia	148
Gymnostcon	65	Manilla Plantain	38
Habrothamnus	145	Maranta	22, 54, 61
Hanging Fern	21	Marcgravia	176
Hatchet Cactus	169	Marguerite	129
Hebeclinium	134	Marsh Mallow	47
Hedgehog Cactus	167	Martinezia	32
Hedychium	69	Masdevallia	82, 89
Hexacentris	147	Maxillaria	92
Hibiscus	47	Melia	130
Himalaya Fern	186, 187	Mesembryanthemum	136
Hippeastrum	111	Mesospinidium	92
Hoffmannia	134	Meyenia	70
Hortensia	114	Micania	148
Hoya	146	Miltonia	92
Hyacinth	104, 110	Mimosa	175
Hydrangea	104, 114	Mimulus	133
Hypolepis	185	Moccasin Plant	83, 96
Illicium	129	Monkey Puzzle	20
Imantophyllum	111	Monstera	20, 21, 47, 58
Impatiens	130	Mormodes	92
Indian Fig Cactus	168	Morning Glory	145
Indian Shot	60	Moth Orchid	94
India Rubber Tree	46	Muehlenbeckia	129
Inga	132	Musa	38
Ipomea	145	Musk	133
Iresine	135	Myriophyllum	159
Isolepis	160	Narcissus	103, 106
Ixia	104, 108	Nelumbium	158
Jasminum	146	Neottia	95
Justicia	47	Nepenthes	172, 173
Kangaroo Apple	145	Nepeta	71
Karatus	68	Nephrodium	188
Kentia	32	Nephrolepis	190
King Cactus	168	Nerium	48
Lady Slipper	82, 83, 96	Nettle Spurge	131
Lælia	82	New Zealand Flax	64
Lapageria	142	Nidularium	68
Lastrea	188	Nierembergia	136
Latania	20, 26, 28	Night-blooming Cereus	165
Laurel	20, 47	Nightshade	145
Laurus	28, 47	Nipple Cactus	166
Lavender Cotton	135	Nymphæa	155

Odontoglossum	92, 93	Pteris	181, 182
Old Man Cactus	166	Ptychosperma	34
Old Man's Beard	132	Queen of the Night	165
Olea	132	Rhapis	26
Oleander	48	Rhododendron	104, 109
Olive	132	Rattan	20, 27, 30
Oncidium	82, 93	Rattle Snake Plantain	95
Onychium	187	Ravenala	39
Opuntia	168	Rhyncospermum	147
Oreodoxa	33	Richardia	58
Osmanthus	132	Rochea falcata	169
Painted Vine	144	Rodriguezia	95
Palmetto	27	Roses	104, 115-117
Palms	23-34	Rubber Tree	21
Palmyra Palm	24	Ruellia	70
Panax	131	Rushes	160
Pancratium	111	Russellia	183
Pandanus	20, 21, 22, 35, 36	Sabal	21, 27
Panicum	65	Saccolabium	95
Paper Plant	160	Sage	133
Papyrus	160	Sago Palm	40
Parrot's Feather	159	Sagittaria	159
Passiflora	144	Salvia	133
Passion Flower	144	Salvinia	160
Paulinia	143	Sanchezia	54, 69
Pelargonium	107	Sanseviera	47
Peleciphorii	169	Santolina	135
Penang Lawyers	26	Sarracenia	174
Peperomia	70	Saxifraga	132
Pepper Vine	143	Scripus	160
Pereskia	169	Schismatoglottis	60
Peristeria	94	Screw Pine	20, 21, 35
Peristrophe	70	Scuticaria	95
Persian Violet	105	Seaforthia	21, 34
Phaius	89	Seaside Grape	44
Phœnicophorum	33	Sedge	160
Philodendron	20, 47, 58	Selaginella	190
Phœnix	33, 34, 52	Sensitive Plant	175
Phorminum	64	Seven Stars	167
Phyllocactus	168	Shield Fern	188
Physianthus	146	Shield Plant	57
Pilea	176	Shingle Plant	176
Pilocereus	166	Silver Thatch Palm	20, 27
Pilumna	94	Sinningia	118
Pineapple	66	Sitilobium	184
Piper	143	Slipperwort	112
Pistol Plant	176	Sobralea	95
Pitcairnia	67	Solanum	20, 140, 145
Pitcher Plant	173	Sophronitis	90
Pittosporum	47	Sparaxis	104, 108
Plantain	38	Sphaerogyne	55, 71
Platycerium	190	Spiderworts	65
Platyclinis	90	Spindle Tree	45
Poinsettia	131	Spleenwort	188
Polypodium	187	Spurge	131
Polystichum	188	Staghorn Fern	190
Pothos	56	Stanhopea	95
Prickly Pear	168	Stenochlaena	187
Primrose	103, 106	Stephanotis	146
Primula	106	Strelitzia	39
Pritchardia	26	Strobilanthus	54, 69

Swainsonia	144	Tulip	103, 106
Sweet Bay	20	Turtle back Plant	71
Sword Fern	188	Tydæa	105, 118
Tacsonia	144	Vanda	80, 90
Talipot Palm	25	Vanilla	83, 95
Tea Plant	129	Verschaffeltia	34
Tecoma	147	Victoria	152, 153, 155
Telegraph Plant	175	Vriesia	67
Thamnopteris	188	Water Hyacinth	159
Theophrasta	21, 48	Water Lily	104, 153, 155
Thunbergia	70, 147	Water Milfoil	159
Thyrsacanthus	70	Water Plantain	159
Thrynax	27	Water Poppy	159
Tillandsia	67	Wax Flower	146
Todea	185	Wax Palm	81
Torenia	133	White Bladder Flowers	146
Touch-Me-Not	130	Woodwardia	187
Toxicophlœa	48	Wandering Jew	65
Trachelospermum	147	Yellow Flax	130
Trachycarpus	34	Zamia	40
Traveler's Tree	21, 39	Zingiber	68
Tree Ferns	182, 183	Zygopetalum	95
Trumpet Creeper	147		

www.ingramcontent.com/pod-product-compliance
Lightning Source LLC
Chambersburg PA
CBHW020832230426
43666CB00007B/1188